Regulation of Vertebrate Limb Regeneration

Regulation of Vertebrate Limb Regeneration

RAYMOND E. SICARD
EDITOR

New York Oxford
OXFORD UNIVERSITY PRESS

Oxford University Press

Oxford New York Toronto
Delhi Bombay Calcutta Madras Karachi
Petaling Jaya Singapore Hong Kong Tokyo
Nairobi Dar es Salaam Cape Town
Melbourne Auckland

and associated companies in
Beirut Berlin Ibadan Nicosia

Published by Oxford University Press, Inc., 200 Madison Avenue,
New York, New York 10016

Library of Congress Cataloging in Publication Data
Main entry under title:
Regulation of vertebrate limb regeneration.
Bibliography: p.
Includes index.
1. Amphibians—Physiology. 2. Extremities (Anatomy)—Regeneration.
I. Sicard, Raymond E.
QL669.2.R44 1985 597.60431 84-25563
ISBN 978-0-19-503604-6

Preface

Interest in and fascination with questions and problems of vertebrate regeneration appear to be increasing. Some of this interest is clearly of a traditional nature, rooted in developmental biology, while other interest seems to be generated from a clinical standpoint, stemming from pathobiology. Consequently, the field is expanding in scope as well as depth. In addition, many students of regeneration have found a progressively greater reliance on techniques and approaches drawn from molecular and cellular biology. This coming of age (of sorts) of facets of regeneration research has caused several problems to be cast in a new light and has required rethinking and reevaluation of traditional questions. While these changes definitely benefit the advancement of our understanding of the process of regeneration, they require that individual investigators narrow their scope, specializing and focusing on particular aspects of the problem.

Although numerous monographs and symposium proceedings on the topic of regeneration have been published over the past several years, the time seems right for this venture. Previous monographs have been the works of single authors and have thus provided a single perspective—one that, although of excellent quality, may not always have been first-hand on the topics discussed. In addition, the several symposium proceedings that have been published have focused on selected aspects and types of regeneration. However, none of these works has addressed directly the topic of the regulation of this process as its primary, or exclusive, concern.

As our understanding of regeneration problems is becoming more sophisticated, an assessment of our understanding of the sourses of this process's regulation seems warranted. To this end, several investigators were asked to provide reviews of the particular aspects of the regulation of amphibian forelimb regeneration on which their research is focused. An up-to-date and expert perspective has been generated for each of the sources of regulatory influence that operates during amphibian forelimb regeneration.

Since no source of regulatory influence functions in isolation, some of the chapters discuss one or another regulatory influence as a subsidiary issue, in addition to the major topic of the chapter. A more holistic and more completely integrated perception of the regulation of amphibian forelimb regeneration can thus emerge. In addition, differences of opinion sometimes occur in describing the significance and impact of certain aspects of this regulation. No attempt has been made to reconcile these apparent conflicts or to impose a uniform interpretation in such cases. Since these areas of controversy arise from a scarcity of information or from the degree of complexity associated with the topic in question, it can only be to the benefit of future studies to allow these issues to surface and the nature of the controversy to be defined. Of necessity, all discussions must remain incomplete and inconclusive. Nevertheless, this monograph will be useful to those embarking on studies of forelimb regeneration and to established investigators with long-standing interest in this area. Moreover, to the extent that other forms of regeneration (e.g., regeneration of liver, muscle, or nervous tissue) share characteristics and features of limb regeneration, the perspectives and information presented in this monograph can provide insights valuable to expanding our understanding of the regulation of these processes as well.

A project such as this is clearly a team effort. It is therefore quite appropriate, although by no means adequate, to express my sincerest gratitude to all who have contributed to its successful completion. The individual authors are thanked for taking the time from their busy schedules to share data and perspectives. I am also indebted to Alexander Wolsky and Bruce Carlson for their encouragement during the early stages of this venture, and to William Curtis and Wendy Warren Keebler of Oxford University Press for their assistance and efforts in guiding this monograph through to publication. For permission to reproduce various figures, I thank the publishers of *American Zoologist* (fig. 7.6), *Developmental Biology* (figs. 7.1, 7.4, 7.5), *Differentiation* (fig. 5.2), the *Journal of Experimental Zoology* (figs. 2.3, 2.4, 2.6, 2.7, 4.2, 4.3), and *Wilhelm Roux's Archives of Developmental Biology* (figs. 4.6, 4.7). In addition, I wish to thank the Purdue Research Foundation for permission to use four original prints produced by Professor David Williams (figs. 2.1, 2.2, 2.5, 2.8). An immeasurable debt of gratitude is also owed to the typists whose skills made the manuscript version of this monograph possible. Lastly, I wish to thank my wife for her patience and forbearance during the many months that this project required.

Framingham, Massachusetts R. E. S.
June 1985

Contents

Contributors

Richard B. Borgens
Department of Anatomy
School of Veterinary Medicine
Purdue University
West Lafayette, IN 47907

Robert L. Carlone
Department of Biological Sciences
Brock University
St. Catherines, Ontario
Canada L2S 3A1

Morton Globus
Department of Biology
University of Waterloo
Waterloo, Ontario
Canada N2I 3G1

Richard A. Liversage
Department of Zoology
University of Toronto
Toronto, Ontario
Canada M5S 1A1

Danielle S. McLaughlin
Department of Zoology
University of Toronto
Toronto, Ontario
Canada M5S 1A1

Hooley M. G. McLaughlin
Department of Zoology
University of Toronto
Toronto, Ontario
Canada M5S 1A1

Anthony L. Mescher
Anatomy Section
Medical Sciences Program
Indiana University
Bloomington, IN 47405

Cherie L. Olsen
Department of Zoology
Miami University
Oxford, OH 45056

Raymond E. Sicard
Department of Pediatrics
Rhode Island Hospital
Providence, RI 02902

David L. Stocum
Department of Genetics & Development
University of Illinois
Urbana, IL 61801

Roy A. Tassava
Department of Zoology
Ohio State University
Columbus, OH 43210

Swani Vethamany-Globus
Department of Biology
University of Waterloo
Waterloo, Ontario
Canada L21 3G1

Regulation of Vertebrate
Limb Regeneration

1 Introduction and Overview

RAYMOND E. SICARD

General Background

Regeneration is a process that has long intrigued students of development. The existence of regenerative capacities was documented as early as the time of Aristotle (*De Generatione Animalium*) and Pliny (*Secundi Historia Mundi*); however, serious interest seems not to have arisen until the middle of the eighteenth century (Morgan, 1901; Vorontsova and Liosner, 1960; Schmidt, 1968). Since that time, the widespread occurrence of regenerative processes has been amply documented (Vorontsova and Liosner, 1960; Schmidt, 1968). More importantly, however, numerous strategies of how plants and animals accomplish regeneration have been described. In particular, these include (1) mechanisms that resemble (or are) modes of asexual reproduction [e.g., similarities to budding and fission among protozoa, coelenterates, and flatworms (Vorontsova and Liosner, 1960)], (2) mechanisms of routine, limited renewal (physiological regeneration) [e.g., the renewal of the lining of the alimentary tract, replacement of circulating blood cells (Vorontsova and Liosner, 1960)] and (3) mechanisms for repair and restoration following acute, traumatic, and severe damage (reparative regeneration) [e.g., liver and kidney regeneration (compensatory hyperplasia and hypertrophy); limb and tail regeneration (epimorphic regeneration) (Vorontsova and Liosner, 1960; Schmidt, 1968)]. [Readers seeking additional general information about regeneration and its phylogenetic distribution are referred to other of the several monographs available for a more detailed discussion (e.g., Morgan, 1901; Needham, 1952; Vorontsova and Liosner, 1960; Hay, 1966b; Goss, 1969).]

Students of development and particularly of regeneration are aware of the changing trends and strategies employed in the investigation of this process. Thus we can recognize an early period during which interest focused primarily on determining whether or not particular organisms possessed the ability to regenerate. This was undoubtedly a major activity for Trembley, Spallanzani, Bonnet, and others during the late eighteenth and early nineteenth centuries (Vorontsova and Liosner, 1960). Overlapping this first period, and extending even to the present, interest and attention were shifted to characterizing the events of regeneration. This second phase can be considered to consist to two interrelated divisions: (1) a chronological documentation of the particular events of the process and (2) the identification of those factors or events upon which the former depends. Significant strides in these areas have been made during the past 75 years using morphological and histological techniques at first, and ultrastructural and biochemical techniques more recently (see Schmidt, 1968, and Wallace, 1981, for good reviews). Extending from the information thus generated, the third phase has emerged. Investigations are now being undertaken in an effort to understand the regulatory processes themselves. This can be discerned by the increasing number of physiologically oriented studies at all levels (systemic, cellular, and molecular) during the past 40 to 50 years.

In one sense, the study of limb regeneration has matured through its descriptive period and is now into a more mechanistic period. Although no one would presume to suggest that the process of limb regeneration or the specific sources of regulatory influence are fully described or characterized, most would acknowledge that we are prepared to seek answers to how the events of limb regeneration are actually regulated. The purpose of this monograph, therefore, is to review our current understanding of the regulation of amphibian forelimb regeneration. We hope that the following discussions will suggest fresh insights into old problems among students of limb regeneration, on the one hand, and bring to light similarities between limb regeneration (and its regulation) and events occurring in other regenerating (or developing) systems, on the other.

Outline of the Events and Sequence of Forelimb Regeneration

The details of the cellular biology of the events of forelimb regeneration can be found outlined and summarized elsewhere (Schmidt, 1968). In addition, detailed staging systems are already published for *Notophthalmus viridescens* (Iten and Bryant, 1973) *Ambystoma maculatum* (Stocum, 1979), and *Ambystoma mexicanum* (Tank et al., 1976). Interested readers are referred to these works for additional details. Therefore no attempt will be made here to present a thorough account of the events of limb regeneration, however, a brief summary of some of the events of this process follows.

Limb regeneration in amphibians requires an initiating stimulus. This is usually in the form of an amputation. The first events, therefore, are those associated with *wound healing*. Characteristically, wound healing is accomplished by first sealing off the wound through migration of epidermal cells, followed by a fairly typical inflammatory reaction. Epidermal closure occurs quite rapidly, being accomplished in a matter of hours, while the inflammatory reaction may persist for several days. The manner in which these events proceed is crucial in determining whether or not regeneration actually occurs. Typically, following amputation, wound healing proceeds without cicatrization, the formation of fibrous tissue deposits; however, should a cicatrix form, all further signs of regeneration fail to appear.

As wound healing progresses, the next significant aspect of regeneration commences. This is the process of *dedifferentiation*. Through the events of dedifferentiation, the phenotypic characteristics of muscle, bone, and other connective tissue are lost by the cells in the immediate wound area as they assume more embryonic-like traits. This loss of specialization, coupled with proliferation, leads to the emergence of a small regeneration bud, or blastema. This marks the end of the preblastemal phase of regeneration.

Entry into the blastemal phase of regeneration signals the initiation of the progressive stages of this process. Over the next several weeks, (1) the blastema enlarges because of increased *cellular proliferation*; (2) the blastema cells become committed to and eventually express once again traits of muscle, cartilage, bone, and other connective tissue through *differentiation* and *histogenesis*; and (3) the form and pattern of the limb are finally restored through *morphogenesis*. Further activity is then directed toward continued growth of the newly formed structures to attain their "adult" dimensions. Regeneration of normal structures, and thus of a normal limb, depends upon each of these events occurring at the proper time and to the appropriate extent.

Influence on the Progress of Forelimb Regeneration: An Overview

The continuum of events that constitutes the process of limb regeneration (i.e., wound healing, dedifferentiation, proliferation, differentiation, histogenesis, morphogenesis, and growth) is subject to influence from one or another source (see Table 1.1). In general, these events are affected by local as well as systemic factors. The most significant of these regulatory influences emanate from the

TABLE 1.1. Phases of Forelimb Regeneration and Sources of Regulatory Influence Affecting Them

Phase	Major events	Sources of regulatory influences active here	Chapters
Preblastemic	Wound healing and demolition	Biopotentials	2
		Skin	3
		Endocrine system	4
		Blood cells/immune system	8
	Dedifferentiation	Skin	3
		Endocrine system	4
		Nervous system	5
		Blood cells/immune system	8
Blastemic	Proliferation and growth	Endocrine system	4, 7
		Nervous system	5–7
		Immune system (?)	8
Morphogenetic	Histogenesis	Endocrine system	7
	Pattern formation	Skin	3
		Immune system (?)	8

endocrine, immune, integumentary, and nervous systems. In addition to the contact-mediated and chemical inputs from these sources, bioelectrical factors are involved in orchestrating the events of limb regeneration.

Following amputation of the forelimb, *bioelectrical potentials* in the limb are altered. Abolition of these particular new potentials in animals that regenerate leads to a suppression of regeneration, while their artificial production in animals that are nonregenerators promotes a regenerative response. These potentials appear to have their origin in the skin (see Chap. 2).

In addition, the *skin* appears to participate in affecting the progress and outcome of regeneration in other ways. Besides specifying the orientation and position of the blastema, the wound epidermis (or apical cap) plays a major role in promoting the formation of the blastema itself. Intimate contact between the wound epidermis and the underlying tissues is required for dedifferentiation. The wound epidermis might also participate with other sources of influence (e.g., neural and endocrine) in promoting proliferation of the dedifferentiated cells. Furthermore, the skin exerts a profound influence on pattern formation during the morphogenetic phase of regeneration (see Chap. 3).

The manner in which wound healing is accomplished is critical to the process of regeneration. Manipulation of the endocrine or nervous system can create conditions that lead to the formation of a cicatricial scar and the subsequent suppression of all further regenerative responses. The behavior of *leukocytes* and of the *immune system* cannot be overlooked in this regard, although these might serve only as targets for neural or endocrine influence (see Chap. 8).

Numerous *endocrine* manipulations have been found to exert varying degrees of influence on wound healing and later stages of regeneration. Although hormones might be affecting regeneration in their role as regulators of metabolic activities, a proper appreciation of their role in regeneration and the mechanisms through which they exert their influence is lacking. It is suggested that a different perception might be required in order to rectify this problem (see Chap. 4).

Neural influence on regeneration appears to be better understood. Without neural influence, a blastema does not form. Thus nerves are important in affecting dedifferentiation. In addition, a major role of nerves in regeneration is that of promoting the proliferation of blastema cells during the progressive stages of regeneration (see Chap. 5).

Significant effort has been made to identify the particular neurotrophic substance(s) responsible for this regulation. Numerous candidates have already emerged, but no final solution to this issue seems imminent (see Chap. 6). However, through these and related efforts in organ culture, a better understanding of the cellular physiology of the blastema cells and a clearer sense of the specific effects of the apical epidermal cap and of particular endocrine and neural products has been obtained (see Chap. 7).

To summarize briefly, during *wound healing*, bioelectrical changes become evident. These in conjunction with leukocyte activities during inflammation might determine whether or not the wound epidermis will interact with the underlying mesodermal tissues. *Dedifferentiation* then results if this interaction has occurred. The events of wound healing and dedifferentiation appear to be influenced by neural and endocrine factors as well. Blastema cell *proliferation* and growth of the blastema, once formed, are affected by neurotrophic factors, a variety of hormones, and perhaps the immune system. Subsequent *redifferentiation* and *histogenesis* occur under similar influences. It is likely that the specific endocrine and neural products that promote redifferentiation and histogenesis differ from those that promote proliferation of the blastema cells. The *pattern formation* then observed during the morphogenetic phase proceeds under the influence of the skin. Lastly, growth to adult dimensions most likely proceeds under endocrine influence (see Table 1.1).

Each of the chapters that follows focuses on one of the sources of regulatory influence on regeneration and elaborates on one or more of the points outlined above.

2 Natural Voltage Gradients and the Generation and Regeneration of Limbs

RICHARD B. BORGENS

In the early 1950s Gordon Marsh and Harold Beams made a striking observation after experimentally manipulating planaria. Using center segments of the animal (in which both the head and tail were amputated), they found they could predetermine which end of the segment would regenerate a head—irrespective of its original polarity. What made Marsh and Beams' experiment intriguing is that they imposed an axial polarity on regenerating center segments of the flatworm using an *applied electric field.* Heads always formed toward the cathode (negative pole) of the applied field. If an original posterior end was aligned in the field such that this end faced the cathode, a two-headed animal was produced, moreover, the effect was dose dependent. With the use of imposed fields on the order of 140 mV/mm, temporary head behavior was noted in correctly aligned tail ends facing the cathode; higher field strengths (180 mV/mm) produced temporary head structures in tail regions; still higher fields (about 200 mV/mm) produced two permanent heads, one on either end of the original center segment.

Results such as these could be taken to suggest that the *natural* controls governing regeneration and growth may have an electrical component. I will discuss the modern evidence linking endogenous electricity to the development and regeneration of limbs in urodeles in particular, and the overall responses to trauma by all vertebrates in general.

Injury Potentials and Wounds

Before the American Civil War it was well known to physiologists that wounding the skin produced endogenous electricity. Early galvanometric studies on wounds were reported by DuBois-Reymond (1843, 1860), Herlitzka (1910), and Monroy (1941). These investigators found that wounds to the skin of a variety of animals (including humans) produced a steady and measurable voltage difference between unwounded skin and the wound. Such potential differences were measured with galvanometers using surface electrodes in contact with the wound and referenced to an unwounded area. All such measurements demonstrated that wounds to the integument were externally electropositive with respect to intact areas of an animal. Since, by convention, the direction of current flow is defined to be that direction in which positive charge would move, we can infer from these data that current *leaves* wounds and probably returns via unwounded areas. Owing to certain technical considerations, the magnitude of such "bioelectric potentials," as well as their physiological significance, is unclear. This has been reviewed elsewhere (Borgens et al., 1979b; Borgens, 1982a); however, briefly, the absolute magnitude of externally measured potential differences depends on the resistance along the animal's body between the recording electrodes. This resistance can fluctuate wildly in aquatic amphibians (e.g., when an aquatic animal is removed from its habitat to an aerial environment for measurement). Since this resistance was not measured (or at least reported), little quantitative information can be gained from these older measurements. What can be gleaned from them, however, is that wounds probably leak current (defined as a flow of ions) to the exterior regions of the integument, and this flow of current is probably associated with a voltage drop, externally positive, in the vicinity of the wound. [For a more detailed explanation of these factors see Borgens (1982a) and Barker et al. (1982)].

The earliest attempt to link such galvanometric measurements to the regeneration of salamander limbs was made by Alberto Monroy (1941). Monroy was the first to study endogenous electricity in amputated and regenerating *Triturus cristatus* limbs. He found that the stump end was electropositive with respect to the undamaged regions, and that the magnitude of these voltages steadily decreased over a period of a month after the limb was severed. Monroy made several different attempts to determine the source of these "translimb" voltages. For example, he removed the heart and checked to see if this affected the character of limb stump potentials. None of his experimental manipulations affected these measurements and he concluded (by process of elimination) that the skin was probably the source of the electrical fields produced by injury along the skin of *Triturus* limb stumps.

Similar measurements have been repeated in modern times by a variety of investigators (Becker 1960, 1961a,b; Rose and Rose, 1974; Lassalle, 1974a,b). Though they used modern high-impedance DC voltmeters to record potentials along the outsides of limbs, the measurements themselves still suffer from the same technical difficulties discussed above because they employed external contact electrodes[1] to record these surface potentials. Furthermore, with regard to *all* such measurements, it is an open question whether voltage gradients on the *surface* of the limb are even pertinent to regenerative events that occur *within* the core tissues of the stump.

All of these investigations, both ancient and modern, agree on one essential detail, however: Wounds are positive with respect to more proximal (uninjured) areas of the body. Thus we should expect the stump to drive current out its end. Moreover, by knowing the magnitude of the current flowing through the core tissue of the stump, we should be able to infer something of the magnitude of the *internal* electric fields associated with this flow of charge.

In view of the technical and procedural uncertainties associated with this older work, a more sophisticated approach was greatly needed. Lionel Jaffe, Joseph Vanable, and I decided to make an attempt using the ultrasensitive vibrating electrode developed in Jaffe's laboratory. This device allows precise and noninvasive measurements of current density moving through a limb with the animal anesthetized and immersed in a natural medium (pond water).

The Ultrasensitive Vibrating Electrode

This device was developed at Purdue in the early 1970s by Jaffe and Richard Nuccitelli. They reported the essentials of its design and construction in 1974. For certain technical details, I direct the interested reader to that paper.

The sensor in this system is a platinum black-tipped electrode (20 μm in diameter) that is vibrated between two measuring positions typically 30 μm apart. The electrode is vibrated by a piezoelectric bender element to which a small voltage is applied. The frequency of this vibration is on the order of 400 Hz. When vibrated in an electrical field produced by a biological (or calibration) current source in a conductive medium, the electrode senses the minute voltage difference that occurs at the extremes of its excursion. This is accomplished by using a phase frequency lock-in amplifier both to set the frequency of vibration and to amplify the signal. With these two modalities tuned together and knowing the resistivity of the medium (simply measured with a conductance bridge), we derive the density of current entering or leaving the source by applying Ohm's law to an extended medium. The vibrating probe system can detect current densities on the order of tens of nanoamperes per square centimeter (depending somewhat on the conductivity of the medium) from current sources as small as 20 μm in diameter.

Salamanders Drive Strong, Steady Current out the Stump's End

Using the vibrating probe system, we detected large densities of current leaving the stump's end in a variety of different salamanders (see Fig. 2.1). Our seminal observations were on the red-spotted newt, *Notophthalmus viridescens* (Borgens et al., 1977a). Since then similar currents have been measured leaving forelimb stumps of the Mexican axolotl (*Ambystoma mexicanum*), tiger salamanders (*Ambystoma tigrinum*), a variety of plethodontid salamanders (*Desomagnathus quadrimaculatus, Batrachoseps attenuatus, Aneides lugubris, Ensatina eschscholtzi xanthoptica*), as well as the West Coast newt (*Taricha torosa*), adult and larval Iberian newts (*Pleurodeles watl*), and the mudpuppy (*Necturus maculosus*) (Borgens et al., 1984).

In all species of salamanders observed, current leaves the stump's end. This is in perfect agreement with the distally positive external voltage gradients as

FIG. 2.1. Ionic current is driven through and out the end of regenerating salamander stumps. The *external* component of this circuit is depicted here (in *Taricha torosa*) as if the animal were in the extended environment of the pond or stream. In semiterrestrial urodeles, the return path of the current would be restricted to the moisture covering the animals. In all urodeles studied, the magnitude of this current flow is on the order of 20–200 $\mu A/cm^2$ soon after amputation, decreases with time at a variable rate (depending on the species), and is driven by the voltage maintained across the skin. (Artist: D. Williams. © 1982 Purdue Research Foundation. Reprinted with permission of the Purdue Research Foundation, courtesy of the School of Veterinary Medicine, Purdue University.)

measured by others. The magnitude of this current ranges from about 20 to over 100 $\mu A/cm^2$ soon after amputation and steadily declines to near base-line values over a period of days. In the red-spotted newt, this decline takes on the order of 10–14 days; for the axolotl, the decline is much more rapid, on the order of 2–4 days. This reduction in current occurs with the increased thickening of the apical epidermis and the first overt appearance of the blastema. Both of these processes occur more rapidly in the axolotl than in the newt. Thus the time course of the fall in current roughly coincides with early events during the process of limb regeneration.

What is the Source of Translimb Currents and Voltages?

Nervous tissue is electrically active and one might suppose nerves to be the source of the endogenous current driven out the stump's end, however, a curious observation was that denervation of the limb appeared, in most instances, to *increase* the current density leaving stumps. A possible explanation for this is the fact that nerves do drive currents of their own; large currents enter the ends of severed nerves (Borgens et al., 1980; discussion in Borgens, 1982a). If current enters the ends of truncated nerves near the surface of the stump, this would decrease the net out-current measured. Thus, if viable nerves are not present in the stump at amputation, one might reasonably expect the density of current to be somewhat higher in denervated stumps when compared to the innervated control. Moreover, this observation strongly suggests that nerves are *not* the electrogenic source of current *leaving* limb stumps (or the source of voltage differences measured along their surfaces). Such a conclusion was also reached by Monroy (1941) and Lassalle (1974a,b) and is suggested by the data of Rose and Rose (1974). Becker has consistently argued that nerves are the source of currents and fields associated with urodele regeneration. Support for this claim apparently comes from three sources: measurements made on an "electronic analog" salamander (Becker, 1960), measurement of transverse "Hall effect" voltage changes before and after denervation (Becker, 1961b), and the apparent loss (or reduction) of externally measured potentials on the limbs of salamanders following section of the brachial plexus (Becker, 1961a). This latter claim has been made without a presentation of the data, without the essential details of the denervation experiments themselves, and without the use of sham-treated controls.[2] Moreover, this claim is at direct odds with the four other independent investigations cited above.

The "electronic analog" salamander had a body fashioned out of sponge, saturated with tap water, containing a soldered wire nervous system. The surface potentials on this sponge salamander were claimed to be similar to a real newt. Also, a discontinuity in the soldered wire nervous system was associated with a loss in the potential gradient, suggesting to Becker that nerves are the source of such surface voltages in real salamanders (Becker, 1960).

Becker's best attempt to show that the nervous system was the source of surface-detected potential differences was by the use of the Hall effect in trying to characterize longitudinal current flow within the limbs of salamanders. (Hall effect voltages are produced at right angles to a flow of current when a magnetic field is imposed perpendicularly to this flow.) It is regrettable, however, that in his brief report the actual electrical records presented are difficult to compare. The recordings showing the Hall effect voltages are at one scale (100 μV/cm) and the recordings of the electrodes alone (without the magnetic field), the effects of brachial plexus section, and section of the cord at the medullae are all presented at another scale (50 μV/cm). One cannot be sure from a close inspection of these records, had they all been presented at the same gain, that any effect of denervation could be easily or reliably delineated. Moreover, there are other inconsistencies in this report. There was a loss of current flow in the *contralateral*, fully innervated limb coincident with denervation of its mate, and a loss of limb potential following sectioning at the medulla, although the major motor nerve plexuses remained intact. These observations are neither explained nor discussed (Becker, 1961b). Lastly, the voltages depended on the "level of anesthesia." This dependence plagues all of Becker's measurements and has not been independently confirmed. We have observed that a variety of anesthesics commonly used on amphibians (tricaine methane sulfonate, urethane, benzocaine) have no effect on the magnitude of the current exiting the stump; that this current does not fluctuate as an animal regains consciousness during any one measurement period; and that it does not behave differently when an animal is immobolized with D-turbocurarine (Borgens et al., 1977a).

Another candidate for the battery driving currents out of limb stumps in salamanders would be the skin surrounding the stump itself. The integument of amphibians is well known to produce substantial voltages across itself (50–100 mV), inside positive. In general, this "skin battery" depends upon an active process of Na^+ uptake from the dilute environment of the pond. The essentials of this skin physiology and its production of transcutaneous voltages are well described (Ussing 1964; Erlij and Ussing, 1978). There are two essential features of this battery that are relevant to our discussion: (1) The polarity of the transcutaneous voltage is inside positive and (2) the production of this voltage, in many but not all cases, is dependent on the concentration of Na^+ in the medium bathing the skin (or at least dependent on the access of Na^+ to the layers of the epidermis that actively pump Na^+ to the interior) (see Kirschner, 1973; Ussing, 1964; Benos et al., 1976).

Monroy (1941) hinted that the skin may be the battery responsible for translimb voltages; however, he had no direct proof of this. Lassalle found that ablation of the skin, altering the Na^+ concentration in the medium, and osmotic shock radically affected the externally measured voltages (Lassalle, 1974b). This was support for the notion that skin is the battery. We found that by raising Na^+ fivefold in the measurement medium (an artificially prepared pond water), the current leaving the stump increased above fivefold. Likewise, we found that replacing Na^+ by choline or simply removing Na^+ from the medium (normally only 1.5–2 mM in Na^+) radically reduced stump currents by about 90% of their original values. The passage of the Na^+ ion through the outermost layers of amphibian skin can be prevented by using several compounds that sterically block the Na^+ channel itself. We have tried several of these agents, dilute in the external measuring medium, and observed their effects on stump currents. Amiloride (Kirschner, 1973; Benos et al., 1976), the methyl ester of lysine, and recently benzamil all reduce stump currents in those salamanders whose skin battery is greatly dependent on Na^+ (see Borgens et al., 1977a, 1984a; Miles and Vanable, 1982). Moreover, these effects are usually reversible. Such tests demonstrate that skin is the major source of current exiting the stump, and the most likely source of potentials measured along the surface of aquatic animals.

Skin Batteries and Stump Currents are Interesting, But are They Relevant to Limb Regeneration?

The fact that the steady decline in stump current coincides with the increasing accumulation of blastemal cells suggests that the two might be related, however, the critical reader would suggest this is weak evidence for the hypothesis. During our search to identify the skin as the active battery, we uncovered a variety of different means to reduce stump currents. We therefore decided to apply these techniques to salamanders and newts and to observe the effects on limb regeneration (see Table 2.1). One would agree that if stump currents played *no* role in the process of limb regeneration, then techniques aimed at reducing the currents should have no effect. Our experiments demonstrated just the opposite: A variety of different applications indeed interrupted the normal regeneration of limbs in salamanders and newts.

Amiloride and Methyl Ester of Lysine Treatments

We applied 0.5 mM amiloride topically to the forelimb stumps of tiger salamanders, beginning at the time of amputation and repeating the treatment every other day for 11 weeks. Control animals were treated with the pond water, which served as the solvent for amiloride. All animals were kept very humid, but not wet, and fed weekly throughout the experiment. Of 33 experimental animals, 48.5% regenerated normally, 15.1% regenerated abnormal and grotesque limbs, and 36.4% were irreversibly inhibited from regenerating. All 34 control animals regenerated normally (Borgens et al., 1979b). Qualitatively similar results were obtained using methyl ester of lysine applications to newts (R. B. Borgens and J. W. Vanable, Jr. unpublished observations).

Since so little of the animals' surface area was treated with these compounds (less than 1%) and since about the same mortality was observed in both groups, we felt confident that a general toxicity was not responsible for these effects.

Regeneration in Na^+ Deficient Pond Water

Tiger salamanders and newts responded similarly to being maintained in an artificially prepared medium deficient in Na^+ (changed daily in order to keep the Na^+-concentration low), but this responsiveness was altogether different from their responses to amiloride.

Newts immersed in Na^+-deficient media showed no external signs of regeneration for 5–6 weeks; most of the control group showed late cone and paddle stage regenerates at this same time. Histological analysis of limb stumps during this prolonged period of inhibition showed the classic signs of arrested regeneration: The stump was covered by full-thickness skin and the end of the bone was covered by callus and connective tissue. On the basis of this appearance, most students of amphibian regeneration would have agreed that limbs should not have regenerated subsequently (see Rose, 1944; Schotté and Wilber, 1958). It took these animals about half the time to reach the same end point (early digits as the control group). We suggested that perhaps some factor(s) was accumulating during this period of inhibition in the experimental group (Borgens et al., 1979c). In urodeles it takes about as long to regenerate a complete forelimb from an amputation through the arm as it does through the forearm. Thus the cues to amputation level induce a more rapid accumulation of blastema and limb outgrowth in the former than in the latter. This positional information can be modulated, as demonstrated by Maden (1982). When a variety of vitamin A analogs were topically applied to axolotls with amputated limbs, regeneration was held in stasis during the application (similar to amiloride and "low-Na^+" treatments discussed above). When the animals were placed back in a normal medium, the limbs regenerated as though a more proximal amputation had been made. Thus striking serial duplications resulted from vitamin A applications.

Perhaps the reduction in current achieved a similar end: After a period of stasis, the limb regenerated, at least *in rate*, as though a more proximal amputation had been made. We observed no serial duplication of structure, however.

Finally, it is reasonable to ask why certain animals in the amiloride series regenerated at all, and why immersion in low-Na^+ medium was ineffective in inhibiting regeneration completely.

With respect to the amiloride treatments, we observed that animals molted irregularly and thus could have cast off the amiloride-treated skin. Urodeles are known to molt irregularly under laboratory conditions and the newly exposed skin is also known to be peculiarly insensitive to amiloride (Nielsen and Tomilson, 1970). Animals may also have washed off the amiloride or diluted it by contacting the moisture condensation on the sides of their plastic containers (Borgens et al., 1979c).

The low-Na^+ treatments were incompletely effective because such immersion may not eradicate the current flow through the stump. Using the vibrating probe system, we spot-checked some animals in this series for current flow and found that stump currents were still being produced (Borgens et al., 1979c). This current could result from the skin's adaptation to the Na^+-free condition and its transportation of other ions to support the transcutaneous battery. We have also learned that divalent cations produce a more sizable contribution to the total ionic flux across amphibian skins than previously suspected, that is, some urodele skin is not completely dependent on Na^+ for the production of the transcutaneous potential. In *A. mexicanum, P.watl*, and *N. maculosus*, it is simply impossible to greatly reduce stump currents by Na^+ modulation. Among other ions, Ca^{2+} is apparently involved in the charge movement across their integument (Borgens et al., 1984).

Recently, Vanable and his group have been developing electrode implants

TABLE 2.1. Electrical Modification of Amphibian Stumps

Species	Treatment	
	Experimental	Control
Notophthalmus viridescens	Chronic immersion in Na^+-depleted APW[a]	Chronic immersion in APW
Ambystoma triginum	Chronic immersion in Na^+-depleted APW	Chronic immersion in APW
Ambystoma tigrinum	Topical application of amiloride to limb stump[d]	Topical application of APW
Rana pipiens (juvenile)	Silver–platinum bimetallic couple[e]	Pieces of silver and platinum wire
Rana pipiens (adult)	Cathode current imposed within limb stump[g]	None used
Rana pipiens (adult)	Cathode current imposed within limb stump: wick electrodes[i]	Sham treatment
Rana pipiens (adult)	Anode current imposed within limb stump: wick electrodes	Sham treatment
Xenopus laevis (adult)	Cathode current imposed within limb stump: metallic electrodes	Sham treatment

[a]Artificially prepared pondwater (APW) was prepared without Na^+. Since Na^+ is very dilute in this medium (about 1.5 mM), there is little change in the ionic strength. Since newt stump currents are dependent on external Na^+, they are reduced by this treatment.

[b]Animals that escaped the initial inhibition regenerated at an accelerated rate and caught up in length and developmental stage with the control population.

[c]Animals maintained in Na^+-depleted medium were checked with the probe system. "Adaptation" currents were stump currents that were independent of Na^+ in the medium.

[d]Amiloride blocks the uptake of Na^+ from the medium by the integument and thus reduces stump currents (see text).

[e]Silver and platinum wire (28 gauge) were heat-fused together and insulated. Such bimetallic couples produce a small voltage between their uninsulated tips when inserted into the animal. The platinum tip is some 0.2 V more positive than the silver tip.

[f]I am in sympathy with Hugh Wallace, who has sharply criticized the use of the term *limb regenerates* induced in adult anurans by a great host of techniques (some described here) (Wallace, 1981). Since the outgrowth happens to arise on limb stumps, I suppose this has been the basis for my own (and others') use of the phrase. However, it is certainly true that no such regeneration ever approaches a limb in external appearance. Even though histological analysis sometimes uncovers a mass of regenerated tissues with various degrees of organization, this is usually not arranged in a manner found in limbs. *Hypomorphic regeneration*, as used in this table, refers to such structures.

Response			
Experimental	Control	Comments	Reference
Limb regeneration inhibited for 5 weeks; then escape[b]	Normal regeneration	No histological evidence for regeneration in animals sacrificed during inhibition	Borgens et al., 1979c
Limb regeneration inhibited for 3 weeks; then escape	Normal regeneration	"Adapation" currents detected with vibrating probe[c]	Borgens et al., 1979c
Regeneration completely inhibited or otherwise interrupted in about 50% of cases	Normal regeneration	Normal regeneration in balance of animals treated with amiloride	Borgens et al., 1979c
Hypomorphic regeneration[f]	Hypomorphic regeneration in few cases	Electrode products likely: similar results with either polarity of implant	Smith, 1967
Hypomorphic regeneration[h]	None used	Electrode products likely	Smith, 1974
Hypomorphic regeneration in all cases	Stump healing only, in all cases	Electrode product contamination unlikely	Borgens et al., 1977b
General tissue degeneration	Stump healing only, in all cases	Electrode products not clearly excluded as effector[j]	Borgens et al., 1977b
External form improved over controls[k]	"Spike" regenerate, typical of adult untreated animal[k]	Electrode products likely	Borgens et al., 1979a

[g]Small hearing-aid batteries (with an appropriate resistor and electrodes) coated with inert polymers were surgically implanted beneath the skin of adult frogs. Current was delivered to the stump tissues with a stainless steel electrode. Such an implant formed the basic design of those discussed below.
[h]Smith claimed to have induced the regeneration of one completely normal limb by these treatments, however, he produced no documentation to properly substantiate this claim (see discussion in Borgens, 1982a, and this text).
[i]Current delivered to body tissues through metal electrodes contaminates these tissues with electrode products. To separate electrical effects from electrode product effects requires the use of salt bridges or aqueous wires (wick electrodes), where current is delivered to the target region by electrolytes.
[j]With the anode wick electrode located at the stump tip, we could not be sure that electrode products (generated at the Ag–AgCl junction) would not have reached the stump (see discussion in text).
[k]Refer to Figures 6 and 7.

that artificially draw current though urodele stumps in the direction opposite to the flow of natural current. The intention is to reduce or eliminate the proximal–distal voltage gradient within the stump. With such a "bucking voltage" imposed within salamander limb stumps, recent trials suggest a substantial reduction in the rate of regeneration and encouraged this group to believe that a more efficient means of reducing (and monitoring) the stumps' internal fields may yield more striking effects (Vanable et al., 1983).

Lastly, it is also satisfying to see how well the hypothesis that endogenous electricity is critical to limb regeneration fits with both the known controlling factors in the regenerative process and the results of other experimental manipulation to limbs (reviewed in Borgens et al., 1979b; Borgens, 1982a).

Endogenous and Applied Currents in Frogs

By now the reader may be somewhat persuaded that the electrical currents and their associated electrical fields might be a factor to be reckoned with in amphibian regeneration. However, on second thought, a peculiar paradox arises: What about grass frogs? The adults of this group (in particular Ranadae) show little if any regenerative ability, yet it is these particular frogs whose skin has been used extensively in characterizing the amphibian skin battery (Ussing, 1964). Thus they would be expected to drive current out their stumps as well.

One clue to this riddle was identified by Meryl Rose (1944). He pointed out a little-known fact buried in a paper published the previous year by Schotté and Harland (1943). These two investigators had carefully searched for changes in the anatomy of tadpole limbs coinciding with their loss of regenerative ability during metamorphosis. Schotté and Harland concluded that there was no obvious change in developmental anatomy that was correlated with this loss of regenerative ability. Rose pointed out that there was at least one: the appearance of the subdermal lymph spaces, which became prominent at the same level of the limb where regenerative ability was lost. Schotté and Harland paid little attention to this detail. However, Rose suggested that because of the lymph space, frog skin may be more loose or "slippery" than urodele skin, and that it might readily advance over the stump, thereby possibly inhibiting regeneration in older tadpoles and adult frogs. Urodele skin tightly adheres to the underlying musculature, and Rose felt that this would happen less in urodeles.

We have reinterpreted Rose's speculation. The large lymph sinuses in frogs can be viewed as a low-resistance pathway (or shunt) for the skin-generated current leaving the stump. If current does preferentially flow out the stump end via this route, then it is likely that the fields within the core tissues of the stump (where regenerative events are taking place) would be greatly reduced.

Using the vibrating probe system, we scanned the stump's end in postmetamorphic frogs (see Fig. 2.2) and found that indeed this lymph sinus shunt

FIG. 2.2. The pathway of ionic current in anuran limb stumps. This stump current is also driven by the "skin battery" (as in urodeles). Its pathway through the limb is quite different, however. Current is shunted through the subdermal lymph spaces found beneath frog skin (but not beneath the skin of salamanders or newts). As depicted here, the largest densities of current leave the periphery of the stump subcircumferentially, with only modest densities traversing the core tissues (not depicted here for clarity). In *Rana pipiens*, current densities leaving the periphery of the stump are two- to fourfold larger than densities leaving the core regions. In urodeles, current peaks were never observed at the periphery of the stump and were always central or postaxial. Moreover, the peak core densities in urodeles were observed to be four- to fivefold larger than in comparable areas of a frog stump (compare Fig. 2.5). (Artist: D. Williams. © 1982 Purdue Research Foundation. Reprinted with permission of the Purdue Research Foundation, courtesy of the School of Veterinary Medicine, Purdue University.)

appeared to operate as we had suspected. Current densities leaving the periphery of the stump were two- to four-fold higher than within the core tissues. Peaks in current density were *never* observed within the central areas of these frog stumps. In all of the newts, current densities leaving core regions of the stump were four- to five-fold higher than at the periphery. Moreover, peaks in current density in these, and nearly all the other hundreds of newt stumps we have scanned, are always central or postaxial (see Borgens et al., 1977a). Therefore there is some reason to believe that differences in limb anatomy between adult frogs and salamanders can help to explain one factor in their obvious differences in regenerative ability (Borgens et al., 1979d).

Additional support for our notion that the necessary densities of current and fields are lacking in frog core tissues (due to the shunt) is the observation that applied electric fields can induce a measure of limb regeneration in adult frogs when imposed on these tissues (Smith, 1974; Borgens et al., 1977b, 1979a).

When small, biologically inert DC stimulators are implanted into some species of adult frogs, with the cathode oriented at the stump's end, regeneration can be initiated (see table 2.1). With the cathode so oriented, current will be pulled through the terminal core tissues of the stump in a proximal–distal direction (see Fig. 2.3). Moreover, the most effective place to locate the cathode at the stump end is in the postaxial region (the area where endogenous current densities are usually greatest in newts) (Smith, 1974; compare with Borgens et al., 1977b).

The abnormal and poorly organized outgrowths resulting from electrical stimulation in *Rana* are certainly not normal limbs, but they are similar to the outgrowths induced in stumps of *Rana* by increased innervation (Singer, 1954) and salt treatments (Rose, 1942, 1944, 1945). However, in these cases, barely postmetamorphic froglets were used as experimental animals. Some of these individuals are known to retain some measure of regenerative ability after metamorphosis (Singer, 1954).

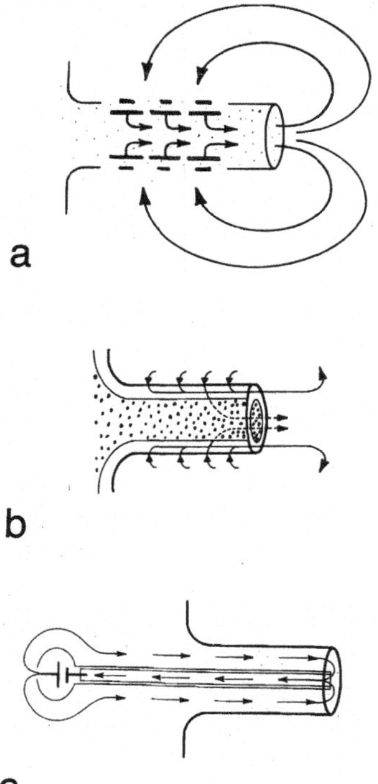

FIG. 2.3. A comparison of various circuits driving current through amphibian stumps. In (a) and (b) the intact skin of the stump drives endogenous charge through and out the amputation surface. (a) In urodeles, this current flow traverses the core tissues of the stump. (b) In anurans most of the skin-driven current flows through the subdermal lymph spaces found in frogs, but not in salamanders or newts (Borgens et al., 1979b,d). (c) An artificially implanted battery and wick electrode (cathode) pulls current through the core tissue of the stump, which is naturally deficient in current (Borgens et al., 1977b, reprinted with permission of Alan R. Liss, Inc.).

Electrical applications, devoid of electrode product contamination, succeeded in initiating a regenerative response in all *large adult* frogs tested (see discussion in Borgens, 1982a; see also Table 2.1 and Fig. 2.4). [The claim by Smith (1974) of one perfect limb regenerate in response to an imposed field is not properly documented and remains unconfirmed (Borgens et al., 1977b, 1979a).] This technique for initiating limb regeneration in frogs is unique in this respect. All of the sham-treated control group stumps healed over with callus and scar. In trials where an anode was located at the stump's end, extensive tissue destruction was observed; however, we could not rule out metal ion contamination in these cases (Borgens et al., 1977b). This is because the positive pole of the battery will produce metal ions and metal ion complexes that would undergo electrophoresis down the salt bridge (wick electrode) in contact with the stump's end. When a negative "wick electrode" is located at the stump tip, any positively charged metallic components liberated at the battery pole would be captured and would not be free to move down the wick.

In another species of adult anuran, the African clawed toad (*Xenopus laevis*), we observed a very peculiar response to artificially applied currents. Normally, adult *Xenopus* stumps produce a hypomorphic limb regenerate that consists of a tapering "spike" of cartilage surrounded by vascular tissue, nerves, and skin. This core of cartilage is unbifurcated and is devoid of nerves in adults. We implanted battery systems similar to those used in *Rana* (Smith, 1974; Borgens et al., 1977b) and pulled current out of the forelimb stump's core tissues (see Fig. 2.5). Over a period of 6–7 months, very interesting limblike structures regenerated (see Fig. 2.6). Early in their development, most regenerates were "paddle" shaped, appearing similar to an analogous stage of urodele regeneration. Many of these developed into curious "pseudolimbs" complete with two to three protruberances. I call these pseudolimbs because histological analysis showed no internal limb differentiation (or other evidence of epimorphic regeneration; see Fig. 2.7). The entire limblike structure was filled with cartilage surrounded by vascular tissue, nerve, and skin. In most ways (save one) the histology was similar to the spikes that regenerated in untreated animals or the sham-treated control group (Borgens et al., 1979a). The fact that some of these pseudolimbs had external features similar to limbs, but without the normal internal components of a limb, challenges how we view the fundamental relationship between developing structure and function (see discussion in Borgens, 1983). Moreover, this implies that endogenous fields may play a role in pattern formation. It is not at all clear how a total applied current of 100 nA delivered for 2½–3 weeks could influence only external form (which took on the order of months to become expressed).

The one major histological difference between the two groups was the presence of large amounts of nerve *within* the cartilage core of electrically stimulated regenerates. Since chondrification is one of the earliest processes occurring in these outgrowths, and since nerves are not known to penetrate cartilage, this suggests the applied current increased both the amount and rate of peripheral nerve regeneration within these limb stumps.

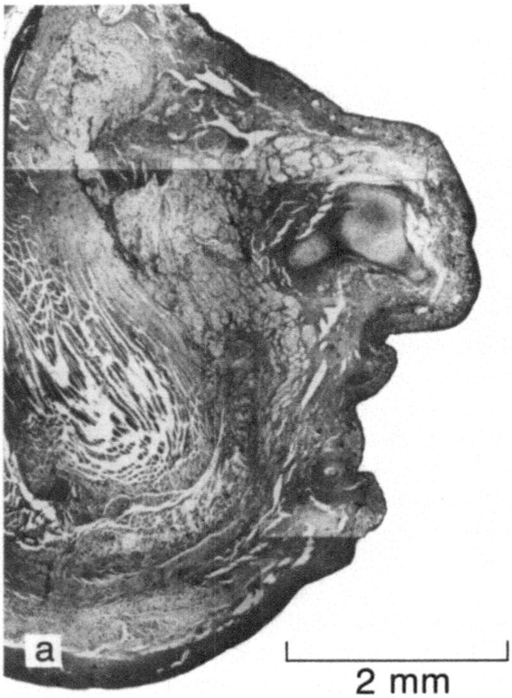

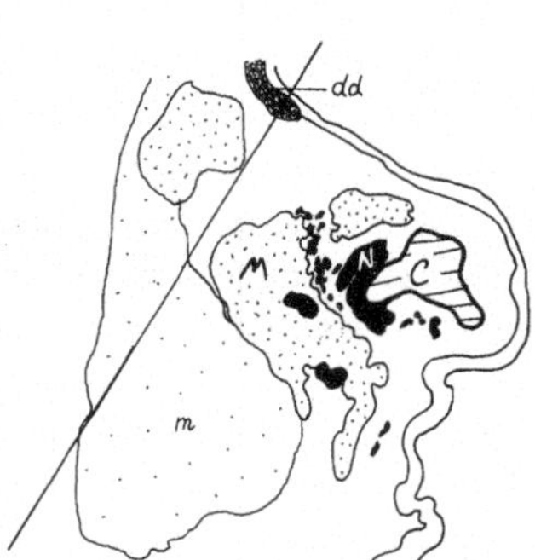

FIG. 2.4. (a) A low-power photomicrograph of a cathodically stimulated limb stump (*Rana pipiens*). (b) Tracing of photomicrograph (M, regenerated muscle; m, original muscle; N, regenerated nerve; C, cartilaginous island; dd, dense dermis; cv, loose connective tissue and vascular tissue). The line approximates the level of amputation. Regenerated bone was observed in other sections. This frog was sacrificed at 10 months, 2 weeks postamputation. In general, regenerates such as this one bore no resemblance to limbs in external appearance. The internal tissues that regenerated showed evidence of a rudimentary organization. All sham-treated animals healed with scar tissue and bone callus beneath a covering of skin and showed no evidence of regeneration. (From Borgens et al., 1977b, reprinted with permission of Alan R. Liss, Inc.)

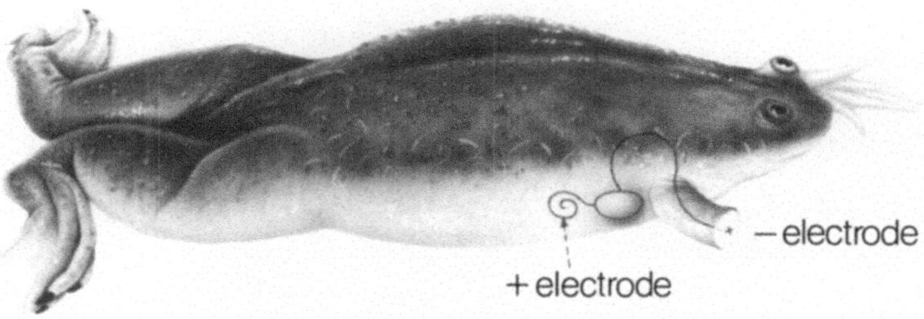

FIG. 2.5. A large adult *X. laevis* implanted with a battery and electrode device. This implantable stimulator was similar to ones used previously in grass frogs (Smith, 1974; Borgens et al., 1977b). The unit was implanted beneath the skin and the negative electrode was routed through the forelimb to make electrical contact with the most distal stump tissues. The stimulator pulled approximately 0.1 μA of current through the limb stump tissues in a proximal–distal direction for 3–4 weeks, at which time it was removed. Forelimb regenerates were analyzed from 6 months to 1 year later. Sham-operated animals were treated identically, except the voltage source was short-circuited internally and delivered no current to limb tissues. The results of these treatments are depicted in Figures 2.6 and 2.7 (see also Table 2.1). (From Borgens et al., 1979a, reprinted with permission of the Purdue Research Foundation, courtesy of the School of Veterinary Medicine, Purdue University. Artist: D. Williams. © 1983 Purdue Research Foundation.)

Cellular Mechanisms and Target Cells

We will discuss whether *currents* or *voltages* are producing biological responses in cells, what magnitude of fields we may expect within limb stump, whether the fields are large enough to be biologically active, and, finally, what may be the target cells.

The physical effect of a current flow through a resistive pathway is heat production. It is implausible that the minute currents discussed here can affect cells through this modality, given the magnitude of these currents and the heat-dissipating properties of tissues and body fluids. Another aspect of current flow that can affect cells is the biochemical action of the ionic components of this flow of charge. This becomes especially pertinent if cells are driving an ionic current through themselves, or if the cell is penetrated by current (either natural or applied). If the ionic composition of cytoplasm is altered in a predictable way, one might expect a cascade of events leading to biological responses.

For example, a local increase in Ca^{2+} conductance into a cell would lead to a localization of Ca^{2+} near its point of entry, since Ca^{2+} is quite immobile and easily bound in cytoplasm. The net result could be a calcium gradient across the cell (see Jaffe et al., 1974). Increases in Na^{+} within the cytoplasm might also produce this effect, since increases in internal Na^{+} could lead to a release of free Ca^{2+} from internal stores such as mitochondria (Carafoli and Crompton, 1976).

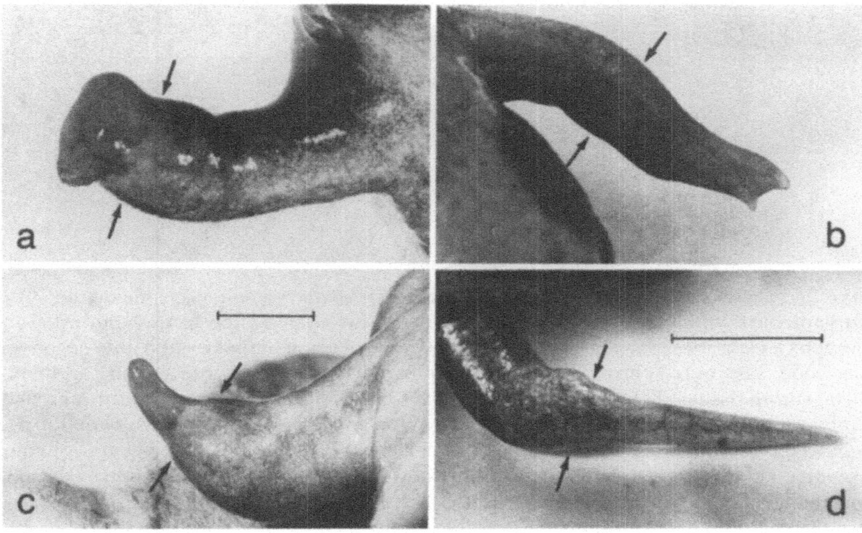

FIG. 2.6. Photographs of *Xenopus* limb regenerates. (a and b) Current-treated limb at 8 and 26 weeks postamputation, respectively. Note the paddle-shaped regenerate in (a), which is indicative of the early stages of normal limb regenerations in urodeles. Variable types of limblike or bifurcated regenerates (b) were observed in response to cathodal current in 6 of 15 animals tested. However, all such limbs (though sometimes striking in external appearance) were hypomorphic in internal structure (see text). (c and d) Sham-treated control limb at 8 and 26 weeks postamputation, respectively. This narrow, tapering "spike" that elongates with development is typical of the normal response to limb amputation in adult *Xenopus*. Such spikes were observed in all control series applications. (From Borgens et al., 1979a, reprinted with permission of Alan R. Liss, Inc.)

Calcium increase (within physiological limits) is implicated in cytoskeletal assembly and dissembly (Schlaepfer and Bunge, 1973; Schlaepfer and Hasler, 1979; Tilney, 1980), cell elongation (Schackman et al., 1978), and a variety of developmental events (see Jaffe, 1980, for an interesting review). Thus the particular ionic component of charge flow through a cell can lead to dramatic consequences, both morphological and developmental.

With respect to most cells found within limbs, it is doubtful whether this hypothesized mechanism is pertinent, with the possible exception of nerves, cells that are extremely elongate, parallel to the long axis of the field, and susceptible to penetration by weak, applied fields (see Jaffe and Nuccitelli, 1977). Most other cells within the path of the current flow are small (~ 10–40 μm) and spherical or stellate in shape (osteoblasts, fibroblasts, chondrocytes, etc.). It is more plausible that this sort of cell would be more susceptible to the

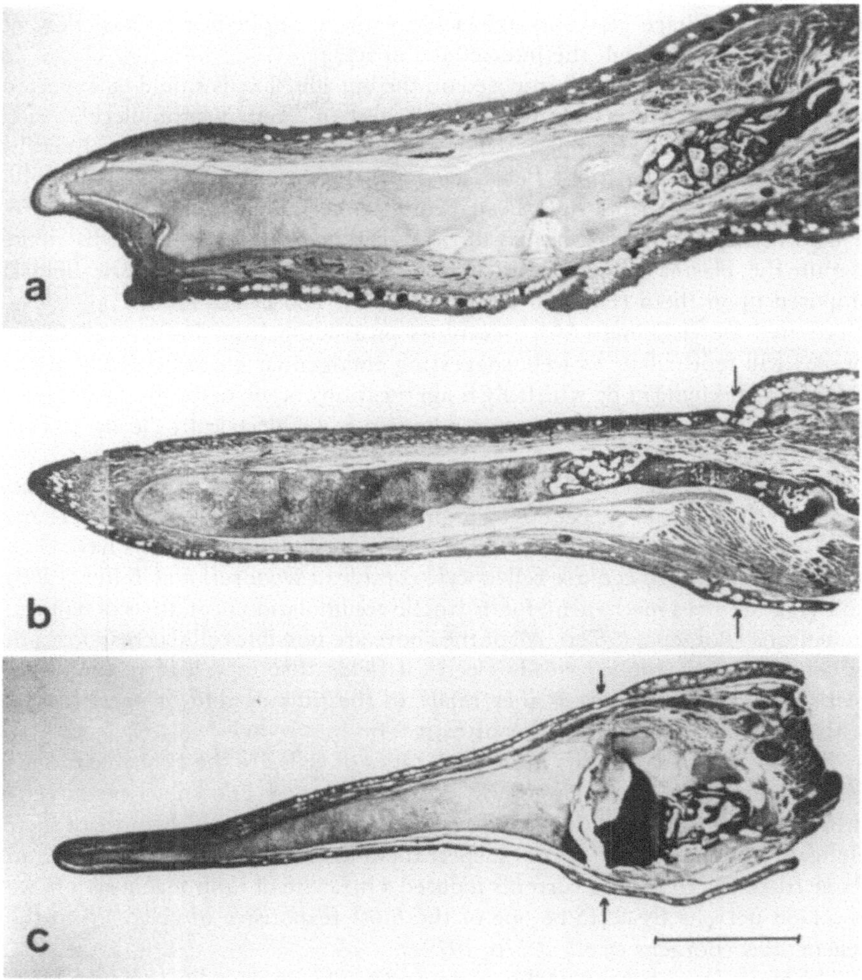

FIG. 2.7. Low-magnification photomicrographs of cathodically stimulated and sham-treated *Xenopus* limbs. (a) Photograph of a cathodically stimulated animal at 11 months and 2 weeks postamputation. The plane of amputation was approximately at the right border of the photograph. Note the unbroken cartilage core partially ossified in its proximal regions. The lacunae contained trunks of nerve fibers. (b) Photograph of another cathodically stimulated animal at 6 months 3 weeks postamputation. Arrows indicate the amputation plane. Note the ossification of the proximal region of the regenerate's cartilage core. Nerve trunks were found throughout this core of cartilage. (c) Photograph of a sham-treated animal at 10 months and 2 weeks postamputation. Arrows indicate the plane of amputation. Note that the cartilage core is not ossified. Nerve trunks were not found to ramify throughout the cartilage core of sham-treated animals. The calibration line represents 2 mm (From Borgens et al., 1979a, reprinted with permission of Alan R. Liss, Inc.)

extracellular voltage gradients associated with an applied or natural flow of current moving through the intercellular spaces.

A steady voltage gradient imposed on the outside of cells would be expected to produce localizations of mobile molecules or even macromolecules suspended in the plasma membrane. (For a theoretical discussion of these hypothesized responses to natural fields see Jaffe, 1977, and Jaffe and Nuccitelli, 1977). Such movements have been proved experimentally in tissue culture. Fluorescently decorated concanavalin A–receptor complexes will redistribute within the plasma membrane of muscle cells when a weak electric field is imposed upon them (Poo and Robinson, 1977; Poo et al., 1978; Poo, 1981). Decorated acetylcholine (ACh) receptors protruding from muscle cell membranes will redistribute as well, suggesting endogenous electrical fields might even be a mechanism by which ACh aggregations occur naturally in the endplate regions of muscle cells (an area known to be electrically "leaky") (Poo, 1981; Betz et al., 1980).

If an applied current penetrates a cell or if a cell drives a current through itself, then one would expect electrophoretic localization of molecules, perhaps even organelles, to occur in response. Such intracellular movements have been studied in the *Cecropia* nurse cell–oocyte complex (Woodruff and Telfer, 1980) and suggested as a mechanism for organelle accumulation near areas of damage in neurons (Borgens, 1982a). All of the above are possible cellular responses to either natural or applied steady electrical fields that may lead to biological responses. These notions, as they relate to the limb system or specific cells within regenerating limbs, await investigation.

We do have some insights into which types of cells are the probable targets of the electric fields within limbs. Nerves are well known to be directly involved in limb regeneration (reviewed by Wallace, 1981) and, because of their elongate morphology, we should suspect them to be ideal targets. Moreover, in those frogs where applied currents induced a measure of limb regeneration, we observed nervous tissue to be one of the most responsive of cells within the regenerates (Borgens et al., 1977b, 1979a).

The epithelium that covers the stump end would also be a likely target. First, this cellular syncytium would be the most resistive element within the path of the current flow, and thus a large voltage drop across this cellular stratum would be expected (than across other cells within the stump). Recent measurements of an increasing resistance across this epithelium with time after amputation suggest this as an important factor in explaining the decrease in stump currents that occurs as wound epithelium becomes established over the end of the stump (J. W. Vanable, Jr., and M. E. McGinnis, personal communication). Finally, the apical epithelium is well known to have a peculiar influence over the accumulation of blastemal cells and perhaps other aspects of the early phases of limb regeneration (see Thornton, 1968; Wallace, 1981).

Do natural stump currents (or the artifically applied currents in frogs) produce a sufficient field within the limb stumps to realistically affect cells? We have suggested that current densities on the average of 20 $\mu A/cm^2$ traversing the stump's core tissues should result in fields on the order of 2–10 mV/mm, given tissue resistivites on the order of 1000–5000 Ω ·cm. Similar densities of current (hence applied fields) should also result from the stimulating units that induced some measure of limb regeneration in frogs (see Borgens et al., 1977b; Borgens, 1982a).

A field of 10 mV/mm is in the low range known to affect various cells in tissue culture. Maximal responses are usually observed in fields 5- to 10-fold larger. Thus the fields within limb stumps are exceedingly weak from a first approximation. Vanable's group has attempted to directly measure the voltage gradients within salamander limb stumps. Although preliminary, their data suggest that the endogenous fields may be on the order of tens of mV/mm. This has led them to favor the wound epithelium as a target for the stump currents, since its native high resistance would ensure a sizable voltage drop across itself. I feel the answer to this dilemma may not come until we can more fully understand the pathway of current between the different types of cells in the limb stump. Since the external electrical field about cells is inversely proportional to the amount of extracellular space the current is flowing through, the degree of packing of cells in various areas of the limb would influence the field strength in that region.

If we add to these complexities the fact that bone produces its own currents of injury (Borgens, 1984a), and nerve its own as well (Borgens et al., 1980), the electrical environment that cells perceive near the stump's end can be very complicated indeed. It strikes me that it may be premature to speculate on what magnitude of field is sufficient for regeneration in the limb system with our present understanding. It is best to simply state that, at least, cells within the limb are responsive to natural and artificially applied currents.

Limb Bud Currents

The relationship between limb regeneration and limb development has long been a stimulating topic of discussion for developmentalists. The recent interest in pattern formation and schemes to model it mathematically have lately given new life to this old paradigm. For a while it seemed that the balance was tipping toward those who believed the two processes were essentially different in mechanism and requirements (see Tickle, 1981; Hinchliffe and Johnson, 1980). However, there are those who feel the pendulum is swinging the other way (see Wallace, 1981; Borgens, 1983). This old debate notwithstanding, it is certain that both processes share great similarities, both in developmental anatomy and physiology. Thus it is reasonable to ask if there are developmental currents associated with limb primogenesis. Only in the last few years has this answer been provided: Currents predict the place of limb emergence in amphibians.

Kenneth Robinson (1983) was the first to detect these limb bud currents. Using a vibrating probe system, he found that current emerges from a general area of hind limb formation in *Xenopus* larvae and that the flux can be detected as early as stage 45 (prebud embryos). The magnitude of these bud currents is on the order of 1–10 $\mu A/cm^2$, and the battery that drives this charge out limb buds is apparently located in the gills. Thus in *Xenopus* larvae there is a large limb bud–gill current loop (Robinson, 1983).

We have confirmed and extended these observations to larval urodeles, where hind limb development takes much longer and occurs in large, more fully formed larvae (Borgens et al., 1983). In the axolotl, currents emerge from a general flank area 7–10 days prior to limb bud condensations (see Fig. 2.8). The locus of peak current density condenses as development proceeds until it is quite restricted to the limb-forming area just rostral to the vent. Current continues to emerge from the elongating bud and finally reverses its direction at the bud's apex, when the bud is on the order of 0.5 mm in length. We have found no evidence of a gill–limb bud current loop in the axolotls similar to that found in *Xenopus* larvae. Instead, regions of in-current at the flank and tail (some distance from the bud) suggest skin is the battery driving charge out the bud in this urodele. Another major distinction from *Xenopus* larvae is the magnitude of the bud currents: Axolotl currents range between 0.5 and 3 $\mu A/cm^2$, two- to fivefold less than found in the African clawed toad. Neither Robinson's group nor ours has had any real success in characterizing the ionic components of this current. Apparently the ionic dependence of these cellular batteries is nonspecific. Thus efforts aimed at replacing, blocking, or preventing any one ion's uptake as a means to substantially lower bud currents is very difficult.

Since the battery driving charge out limb buds is not easily manipulated, it will prove difficult to chronically interfere with it and, in so doing, test its

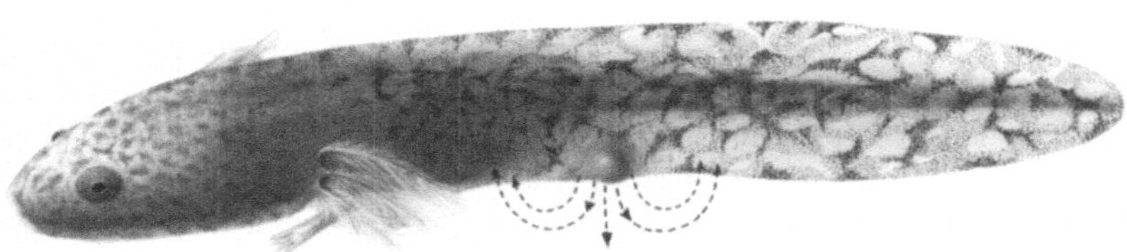

FIG. 2.8. An immature axolotl drives current out of limb-forming areas of flank. Current (about 1-3 $\mu A/cm^2$) emerges from this area on the order of a week prior to limb bud formation. The current localizes over the discrete area of limb formation with development and decreases in intensity after bud formation and growth. Similar limb bud currents have been described in larval *Xenopus* (Borgens et al., 1983; Robinson, 1983). (Artist: D. Williams. © 1983 Purdue Research Foundation. Reprinted with permission of the Purdue Research Foundation, courtesy of the School of Veterinary Medicine, Purdue University.)

relevance to limb formation. Some other clever approaches are needed here to more directly approach the question of relevance. Perhaps indwelling electrodes could be used to artificially accentuate or detract from this natural current.

The observation that current leaves both regenerating limbs and the developing limb bud is in itself satisfying. Given a skin battery, one would expect current to be driven out from the surface of a limb stump after amputation, from wounds, or from other traumatized regions, since in all these instances there would be a current path of very low resistance produced by surgical intervention. In the developing bud, however, there is no trauma to produce the electrical leaks that result in the peaks of out-current prevailing over presumptive limb tissue. Instead, the flank epidermis becomes electrically leaky because of developmental changes in these areas, perhaps owing to changes in the degree of sealing of the epithelial cells in this region (see Borgens, 1984b). In a sense, then, the evidence for the relevance of electric fields in limb regeneration lends support to the relevance of the normal limb bud currents, and vice versa.

Contradictions and Qualifications

Recently Lassalle has argued that bioelectrical factors are unimportant to regeneration in urodele limbs (Lassalle, 1979, 1980). This had been discussed elsewhere (Borgens, 1982a; Borgens et al., 1984), but it would be worthwhile to briefly touch on Lassalle's argument.

Lassalle was unable to measure potentials along the surface of axolotl or larval *Pleurodeles* stumps and claimed that these animals do not produce injury potentials until the adult form (experimentally induced to develop in the case of the axolotl). Axolotls and *Pleurodeles* are excellent regenerators.

I have already touched on the capricious nature of surface-detected potential measurements. Lassalle's inability to measure potentials in the larval (or neotenous) forms is a good case in point. Axolotls certainly do possess stump currents (Vanable et al., 1983; Borgens et al., 1984); moreover the skin battery is probably ubiquitous in amphibians (Borgens, 1982a; Borgens et al., 1983, 1984). Lassalle's problem was almost certainly a technical inability to measure the shallow voltage gradients that occur along the skin surface in these animals. Lassalle also suggested that larval amphibians do not possess a transcutaneous battery. In more recent measurements, inwardly positive voltages have been recorded across the skins of larval amphibians (Cox and Alvarado, 1979; McCaig and Robinson, 1982).

Lassalle claimed that limb stump potentials disappear during a molting period, and thus they cannot be pertinent to limb regeneration. An increase in the surface conductance of skin is a well-known aspect of the molt (reviewed by Larsen, 1976). Such an increase in conductance would be associated with a decrease in resistivity, since the two are inversely proportional. This decrease in resistivity would make surface potentials even more difficult to detect (all investigators employing this method seek to increase the resistance between the surface electrodes to achieve a measurable voltage).

Lassalle was struck by the fact that positive voltages are found near stumps as well as near simple skin wounds. He felt that this suggests such potentials cannot then be relevant to both processes. One might stand his argument on its head and as easily say that the presence of a qualitatively similar endogenous flow of current in all wounds to the integument suggest a universal (or evolutionary) relevance. In fact, the best direct and circumstantial evidence seems to suggest that this notion is probably reliable (refer to an earlier discussion, Borgens, 1982a; Barker et al., 1982; Borgens et al., 1984).

We do not yet have unequivocally clear tests in either limb regeneration or skin wound healing that allow a definitive statement on the relevance of endogenous current to these processes. We certainly need some clever and direct tests to ferret out these answers. For example, if one modulates the voltage across or along skin, will one modulate the character or rate of skin wound healing? Can one electrically induce limbs to form in animals where they would not naturally occur? This could be attempted using artificial applications of current to the legless flank regions of sirens or to the embryos of snakes.

There are other problems to tackle before we gain a better understanding of current flow in urodele limbs. A more precise understanding of the *field strengths* inside limbs would aid our quest for the target cell of endogenous (or artificially applied) electric fields. At present we have only inferences based on measured current flow through the core tissues of the stumps. We also need a more precise knowledge of the way bone and nerve (and perhaps certain other tissues in limbs) produce their own endogenous currents in response to injury (Borgens, 1984a), and how they may (or may not) respond to them. These tissues in particular are apparently responsive to fields, are major elements within the pathway of the current, and may influence the overall electrical character of the amputation.

Mammalian Regeneration

Anyone who has ever watched a salamander grow a new leg, much less research the controls of such a remarkable process, has certainly wondered about the possibility of initiating (or restoring?) such regeneration in humans. In fact, many experimental treatments derived from amphibian studies have been applied to mammals: salt treatments to rodent digits (Neufeld, 1980), after the fashion of Rose (1942, 1944, 1945); increasing the amounts of nervous tissue in the opossum and rat (Mizell, 1968; Bar-Maor and Gitlin, 1961), after the fashion of Singer (1954, 1965); and increasing levels of various hormones—notably glucocorticoids—after amputation of rodent digits (Schotté and Smith, 1961), fashioned after the notions of Schotté and Wilber (1958). This list is meant to be illustrative and not comprehensive; however, in all such experiments the degree of regeneration achieved has been either nonexistent or unimpressive.[3] The claims made by these investigators have been quite conservative and are not very encouraging to the notion that significant regeneration in mammals can be achieved.

Careful and measured claims (such as those of Neufeld, 1980) are in direct contrast to reports suggesting that "partial limb regeneration" may be induced in rats by applications of electric current to their stumps (after the fashions of Smith, 1974, and Borgens et al., 1977b, 1979a). I have reviewed the details of this controversy elsewhere (Borgens, 1982a); suffice it to state here that these investigators displayed no tissue arising after amputation that would not have arisen anyway in untreated rats, or which could not clearly be excluded as a residue of the amputation (partially cytolyzed bone chips perhaps mistaken as anlangen). I suppose one is somewhat protected by the use of the phrase "partial limb regeneration"—which certainly lends itself well to subjectivity. One person's "partial limb" may well be fracture callus to another. Semantics aside, the evidence presented within these very reports suggests that imposed electricity induced nothing that did not form in the control groups (compare Becker, 1972, with Becker and Spadaro, 1972, referring to discussion in Borgens, 1982a). That the response to current in rat limb amputations could not be separated from "innate" responses was admitted by Libbin and co-workers (1979). Thus will rats "partially" regenerate their limbs normally after amputation?

From the foregoing discussion one would suspect that mammalian limbs, even digits, are completely devoid of regenerative ability; however, such is not exactly the case. Children's fingers (Illingworth, 1974) and the tips of mouse digits (Borgens, 1982b) regenerate after amputation *if* the plane of amputation passes distal to the last interphalangeal joint. This regeneration is not the same as the epimorphic regeneration of the amphibian limb (see Borgens, 1982b); however, they share several interesting similarities (besides the fact that a perfect replacement part is produced in both instances).

In both mammalian digits and urodele limbs, a skin flap covering the stump surface can inhibit regeneration (Illingworth, 1974; Mescher, 1976). A strong endogenous current is driven through the core tissues and out the stump's end in salamanders, children (Illingworth and Barker, 1978), and mice (R. B. Borgens, unpublished measurements). Moreover, the magnitude and character of these stump currents is similar in all three cases. One might reasonably suggest that the reason why skin flaps inhibit regeneration is because the high-impedance skin (Barker et al., 1982) might block the flow of current out the stump. This flow of injury current in the digits of mammals has never been tested for relevance, so it is wise at this juncture not to assume that the process of fingertip regrowth and endogenous current flow are related.

Though the battery driving charge out mammalian digits has not been described experimentally, it is nevertheless likely that it is a "skin battery" similar in most respects to that of amphibians. Though not well known to biologists, the skin of most animals, including mammals, supports an inwardly positive potential difference across itself (Barker et al., 1982, and discussion in Borgens, 1982a). It is the short-circuited flow of current produced by this integumentary battery that is observed leaving lesions in animals.

This current by itself cannot be the major factor explaining the regrowth of digits, since an amputation proximal to the interphalangeal joint in mammals does not result in regeneration, yet current flows out these amputations as well (R. B. Borgens, unpublished measurements on mice). It will be extremely interesting when one uncovers the critical difference between tissues of the distal phalanx (which regenerate) and the same types of tissues just 200 μm more proximal (in the mouse) that do not regenerate (see Borgens, 1982b). Within the digit there must be some curious difference between distal tissues and proximal tissues in their responsiveness to signals induced by trauma. Perhaps one of these "signals" may be electrical in nature.

Summary

A strong skin-driven current leaves the stumps of regenerating salamander limbs, traversing the core tissues of the stump. Inhibiting or reducing this flow of current adversely affects limb regeneration.

Adult grass frogs, whose limbs do not regenerate, also produce stump currents. Peaks in these currents are adjacent to the subdermal lymph spaces in frogs and are always central or postaxial in salamanders. Thus endogenous currents and fields are much reduced within the core tissues of frog stumps when compared to salamanders.

An endogenous current is driven out presumptive limb regions in larval *Xenopus* and the axolotl. This current is localized over the emerging bud and can be found to predict the location of limb bud placode formation days before its emergence in the axolotl.

The production of steady currents of injury and their associated fields are a common feature of animal integuments. This physiological flow of charge may be one of the earliest signals to trauma, both at the time of the acute injury and in an evolutionary sense.

Acknowledgments

I gratefully acknowledge the critical reading and numerous suggestions made by my colleagues Joseph Vanable, Jr., Robyn Overall, and Richmond Prehn and the exceptional artwork by David Williams. I also appreciate the financial support of the National Institutes of Health, the National Spinal Cord Injury Foundation, and the Paralyzed Veterans of America during the years that many of these experiments and ideas were being developed.

Notes

1. Paradoxically, measurements made with galvanometers are sometimes easier to interpret than those using modern high-impedance voltmeters. One can actually infer some of the current densities associated with these measurements, since much of the endogenous current could traverse the galvanometer itself (owing to its low input resistance). We contacted Monroy (Borgens et al., 1977a) and found that his galvanometer (as used in the 1930s!) had a resistance of 1 kΩ. We calculated that his voltages of about 15 mV (at day 1) would be associated with a current density of about 15 $\mu A/cm^2$ leaving a 1-mm^2 limb stump. This estimate would be low, considering that not all of the endogenous current would be expected to traverse the galvanometer. Thus a reasonable estimate would be higher, and within the range detected by our vibrating probe system some 40 years later.

2. This is especially important, since the usual surgical approach to the brachial plexus is through the shoulder region of a urodele. These surgical lesions to the skin would be interposed between the external contact electrodes, one of which is usually placed on the trunk of the animal's body, with the other at the stump's end. For the measurement of surface potentials, a proper control group would be to make a surgical approach to the plexus but not sever it. One would compare surface potentials in this sham-treated group to the denervated experimental group.

3. The often cited example of regeneration in neonatal opposum by Mizell (1968) has recently been questioned by Tassava's group (Tassava and Fleming, 1981). As it turns out, separating the *regeneration* of limb components from the *development* of limb components in the marsupial is a tricky chore, and Mizell's original observations may perhaps be flawed.

3 Role of the Skin in Urodele Limb Regeneration

DAVID L. STOCUM

The skin serves a number of functions in vertebrates, such as (1) protection from dessication, invasion by pathological microorganisms, and minor mechanical trauma; (2) temperature regulation; (3) sensing temperature, pressure, and noxious stimuli; and (4) respiration. The subject of this chapter is to discuss another, most unique function of the skin in urodele amphibians: its role in the regeneration of limbs. Larval and adult urodeles (salamanders and newts) are noted for their ability to regenerate limbs and tails, and the skin of these appendages, especially the epidermis, is crucial to this process.

Structure of Adult Amphibian Skin

Adult amphibian skin, like that of other vertebrates, consists of two components, the *epidermis* and the *dermis*. The epidermis consists of several layers of epidermal cells. The outer layer of flattened cells, the *stratum corneum*, is keratinized. The inner layer of columnar or cuboidal cells, which rests on a prominent *basement membrane*, is the *stratum germinativum*. The intermediate cells are polyhedral in shape. The outer layer of epidermis is shed regularly, and the shed cells are continually replaced by division of cells in the stratum germinativum and intermediate layers. The epidermal cells are held together on their lateral borders by frequent desmosomes and numerous interdigitating microvilli, and the basal surface of the stratum germinativum cells is attached to the basement membrane by hemidesmosomes. The lateral and basal surfaces of the epidermal cells stain with ruthenium red, indicating they are coated with glycosaminoglycans (Repesh and Oberpriller, 1980). The basement membrane consists of two components, a *basal lamina* (adepidermal membrane), which lies directly beneath the stratum germinativum, and a subjacent *reticular lamina* (adepidermal reticulum) composed of numerous sheets of collagen fibers crisscrossing one another (Singer and Andrews, 1956; Salpeter and Singer, 1959, 1960a; Overton, 1976). Within each sheet the fibers are thickened at regular intervals to form nodes (Salpeter and Singer, 1959; Kemp, 1961), and in regions where fiber direction changes substantially, three- and five-sided regions are outlined by the intersecting fibers (Overton, 1976). In addition to collagen, the basement membrane has been shown to contain fibronectin, reticulin, and laminin (Repesh et al., 1981; Gulati et al., 1983).

The *dermis* consists of two layers. The first is the *stratum spongiosum*, a loose network of fibroblasts and pigment cells beneath the basement membrane. The pigment cells extend processes through the basement membrane into the epidermis. In addition, the stratum spongiosum contains large glands, which are derived from the epidermis. The ducts of these glands are narrow and open to the surface of the epidermis, and the basement membrane follows the contours of the glands where they are present (Andrews and Hickman, 1974). The stratum spongiosum is rich in capillaries and lymph spaces and is traversed by many sensory and autonomic nerves, branches of which penetrate the epidermis. Below the stratum spongiosum is the *stratum compactum*, consisting of wavy sheets of collagen fibers, between which reside fibroblasts. Between the dermis and the muscles there is a subcutaneous layer of connective tissue, rich in pigment cells, blood vessels, and lymph spaces. Fibronectin is present in the connective tissue matrix of the dermis, but neither reticulin nor laminin have been detected (Repesh et al., 1981; Gulati et al., 1983).

The structure of the skin varies in different parts of the body. For example, in the axolotl, there is a well-developed dermis in the skin dorsal to the hindlimb, but not in the skin of the tail (Hofmann et al., 1978).

Development of Larval Skin

It is important to recognize that the structure of adult skin is the result of a developmental process and that skin structure therefore differs between larva and adult and at different larval stages. Since different investigators use animals of different ages for experiments on the role of skin in regeneration, such structural differences could influence the kinds of results obtained and their interpretations. The developmental changes in larval skin have been studied histologically in urodeles by Stearner (1946) and ultrastructurally in frog tadpoles by Kemp (1961). The following summary is derived largely from their work, with other studies, mentioned below, providing additional data.

In recently hatched larvae, the epidermis is thin and made up of epidermal cells interspersed with large Leydig cells. The function of these cells is still uncertain (Rosenberg et al., 1982), but they may secrete an interepidermal mucus that confers protection from viruses and bacteria (Andrews and Hickman, 1974). Discrete epidermal layers are not readily recognized and the outer epidermal cells are not keratinized. The epidermis is underlain by a thin basement membrane, but there is no cellular dermis as such at this stage. The dermis will be derived from stellate mesenchymal cells and melanocytes that are closely apposed to the underside of the reticular lamina.

By the mid-larval period, the epidermis has thickened into several cell layers. In metamorphosing species, the Leydig cells decline in number by a process of degenerative hypertrophy and removal by macrophage-like cells; by metamorphosis, Leydig cells have completely disappeared (Andrews and Hickman, 1974; Rosenberg et al., 1982). The reticular lamina becomes thicker (i.e., is composed of several plies), and many of the mesenchymal cells beneath it assume a rounded appearance. Both the mesenchymal cells and the melanocytes now begin to invade the reticular lamina. As the larva grows, these cells make their way between the collagen fiber sheets of the reticular lamina and multiply there. As the larva approaches metamorphosis, the reticular lamina splits, creating a space occupied by a loose meshwork of fibroblasts and melanophores. This space is the stratum spongiosum. The dermal glands begin as nests of basal epidermal cells that sink into the stratum spongiosum and subsequently differentiate as gland cells (Holder and Glade, 1984). These glands are thought to secrete mucus or poison (see Andrews and Hickman, 1974), although there is some uncertainty in regard to their function, at least in axolotls (Holder and Glade, 1984). The thin acellular layer of reticular lamina above the stratum spongiosum becomes the reticular lamina of the adult animal, while the cellulated remainder of the reticular lamina below the stratum spongiosum becomes the stratum compactum. The collagen sheets of the reticular lamina, which earlier were flat, develop their nodular thickenings and are thrown into folds that follow the foldings in the epidermis. The larval collagen sheets are orthogonal to one another, but the angle of the sheets to one another becomes less than 90° at metamorphosis.

Regenerative Response of the Skin

Amputation of a larval or adult urodele limb evokes an epimorphic regenerative response. After wound healing, the mesodermal tissues at the tip of the limb stump become disorganized and their various specialized cell types *dedifferentiate* to a common mesenchymal cell morphology. These cells are termed *blastema cells* and they accumulate and proliferate under the wound epidermis to form the *regeneration blastema.* The blastema grows and *redifferentiates* the missing limb parts in continuity with the parent limb tissues. Redifferentiation takes place in a proximal to distal and anterior to posterior sequence (Stocum, 1979). Here we shall examine the regenerative response of the skin following amputation.

Epidermal Wound Closure

Most studies on epidermal wound closure have been done on adult newts (see Singer and Salpeter, 1961; Schmidt, 1968). The most detailed account is the ultrastructural studies by Repesh and Oberpriller (1978, 1980), and the following summary description is based on their work. After amputation, a fibrous plasma clot forms immediately over the wound surface. Within 2 hr, basal cells at the cut edge of the epidermis mobilize in preparation for migration over the clot. Signs of this mobilization are the rounding up of the cells, the disappearance of hemidesmosomes, detachment from the basal lamina, and reduction of microvillar and desmosomal contacts. A single cell layer at the leading edge of the wound epidermis advances over the clot by extension of pseudopodia, often trailed by a front of cells three to four layers thick. The sheet apparently moves by adhesion of the pseudopodia of the leading cells to the fibrin substratum, with more proximal cells then rolling or sliding over the lead cells to become lead cells themselves. The edge of the migrating sheet meets in the center of the wound, where it piles up in a ridge or ridges, and wound closure is complete within 9 hr.

Epidermal cell migration is not inhibited in skin wounds of adult newts by either colchicine or cycloheximide (Dunlap and Donaldson, 1978; Donaldson and Mason, 1978). Hence neither protein synthesis nor existing microtubules are required for the process, even though migrating epidermal cells exhibit several morphological trademarks of protein synthesis, such as conversion from a condensed heterochromatic nucleus to a dispersed euchromatic one, prominent nucleoli, and abundant free and membrane-associated ribosomes (Repesh and Oberpriller, 1980). Concanavalin A inhibits epidermal migration (Donaldson and Mason, 1977, 1978). The mechanism of this inhibition is unknown, but must involve the binding of the lectin to glycoproteins or glycolipids of the cell surface coat that may be important for adhesion to the fibrin substratum.

Development of the Wound Epidermis During Blastema Formation and Growth

Epidermal wound closure takes place in the absence of cell division, the epidermis at the cut edge of the skin providing the necessary cells by thinning out for a short distance proximally (Hay and Fischman, 1961). There is little cell division in either the wound or stump epidermis during dedifferentiation, but the wound epidermis increases in thickness at the center of the amputation surface by the piling up of distally migrating cells to form an *apical epidermal cap*, which is especially prominent in the axolotl (Tank et al., 1976). As blastema cells accumulate at the tip of the limb stump under the wound epidermis, DNA synthesis and mitotic activity increase in both the wound epidermis and in the stump epidermis adjacent to the blastema–stump junction, but are always higher in the latter (Hay and Fischman, 1961). As the blastema forms and grows, cell division increases in the epidermis of the blastema and adjacent stump and then decreases again during redifferentiation.

Although ducted glands are not found in the blastemal epidermis of adult newts, structures characteristic of normal skin epidermal cells, such as mucus granules and tonofilaments, are present (Repesh and Oberpriller, 1980), and

Leydig cells are observed in the blastemal epidermis of larval limbs (Stocum, 1979). Hence, even though structural alterations in epidermal cells are necessary for the migratory movements of wound closure, these cells nevertheless appear to maintain some characteristics of their original differentiated state. This state is completely restored as the blastema redifferentiates.

One of the major differences between regenerating amphibian limbs and other vertebrate limbs that merely undergo wound healing after amputation is that in the amphibian limb a basement membrane is not immediately reconstituted under the wound epidermis (Salpeter and Singer, 1960a) and the granulation tissue typical of scarring does not form (Schmidt, 1968). The basal lamina reappears at the base of the blastema as the latter forms. As the blastema grows, the basal lamina is reconstituted in a proximal to distal direction, but does not reform at the apex of the blastema until redifferentiation has set in. The reticular lamina reappears under the basal lamina in the early bud blastema and is also reconstituted in a proximal to distal direction, but remains absent apically until the digital buds have appeared (Salpeter and Singer, 1960a). Thus the mesenchymal cells of the blastema remain completely open to communication with the apical epidermis until redifferentiation begins and are separated from the epidermis by only the thin basal lamina until fingerbud formation. Redifferentiation of the dermis presumably follows its original developmental sequence and attains the structural complexity appropriate to the age of the animal, although this has not, to my knowledge, been studied. (For an additional discussion of the role of the apical epidermal cap, see Chap. 7.)

Contribution of Skin to the Blastemal Mesenchyme

Dermis

All of the mesodermal tissues of the limb contribute to the formation of the regeneration blastema. It is currently held that of the two major skin components, epidermis and dermis, only the dermis contributes to the blastemal mesenchyme. This assumption will be followed here for purposes of discussion, and the evidence for and against an epidermal contribution will be discussed in the following section.

By comparing the number of dividing cells in the various dedifferentiating tissues of amputated adult *Notophthalmus viridescens* limbs, Chalkley (1954, 1959) concluded that connective tissue (including cartilage) contributed most of the mesenchymal cells to the blastema. The contribution of the dermis itself was not determined, but the subepithelial connective tissue, which included the dermis and subcutaneous connective tissue layer, accounted for 3.4% of the mitoses. This is the only study that has attempted to quantify the contributions of various tissues to the regeneration blastema, and it suffers from the inherent imprecision of determining the actual tissue origin of a cell that is dedifferentiated and dividing. The contribution of dermis and other tissues might be more precisely assessed by making chimeric limbs of diploid and polyploid tissues and then determining the percentage of cells in the blastema derived from the donor and host tissues. Furthermore, it must be recognized that the dermal contribution to the blastema may vary depending on the developmental age of the animal, since the cellular component of the dermis is at first lacking.

A major question is whether blastema cells derived from the dermis (or any of the mesodermal limb tissues) redifferentiate only according to origin during regeneration, or whether they can undergo metaplasia to other cell types. It is known that the cartilage of a regenerate is not all derived from pre-existing cartilage during normal regeneration (Steen, 1968, 1970; Patrick and Briggs, 1964); hence some must come from other connective tissue or muscle. A metaplastic conversion of dermal cells to cartilage and other cell types has been demonstrated in cell-marking experiments where the ability of non-skin tissues to participate in regeneration was inhibited by x-irradiation. When triploid skin (but not when just triploid epidermis) was grafted to diploid axolotl limbs that were irradiated and stripped of their skin, regenerates with a normal skeleton and other connective tissues, but lacking muscle, were formed after amputation (Namenwirth, 1974; Dunis and Namenwirth, 1977). Most of the mesodermal cells in these regenerates were triploid, demonstrating their origin from the donor dermis.

Transformation of muscle into cartilage and other connective tissue has also been claimed during regeneration of *Xenopus* tadpoles after grafting clones of cultured myoblasts, traceable by means of a nucleolar marker, into hindlimbs (Steen, 1973). A similar transformation has been reported for clones of late embryonic rat and chick myoblasts cultured on demineralized bone matrix (Nathanson and Hay, 1980a,b).

Cartilage can probably undergo metaplasia into other connective tissue, but its transformation into muscle is questionable. Scapular cartilage implanted into irradiated adult newt limbs formed a blastema that subsequently redifferentiated only into irregular masses of cartilage (Eggert, 1966). Cartilage from white or dark axolotls homografted into x-irradiated limbs of opposite color allowed the production of donor-derived regenerates with cartilage and connective tissue (Wallace et al., 1974; Maden and Wallace, 1975). These authors also reported the presence of muscle in these regenerates, but no cross-striations are visible in the tissue illustrated; it is thus probable that this tissue is tendon. Trampusch and Harrebomée (1969) also reported that femoral cartilages implanted under a wound epidermis on the flank of axolotls dedifferentiated and gave rise to regenerates with muscles as well as skeleton, but there was differentiation of flank muscles in these experiments; hence blastema cells derived from flank muscle might have given rise to the regenerate muscle. On the other hand, Desselle and Gontcharoff (1978), using cytophotometric measurements to distinguish between graft and host cells, claimed that blastema cells derived from *Desmognathus fuscus* limb cartilage implanted into irradiated limbs of *Triturus cristatus* formed muscle cells during regeneration of the limbs.

The most appropriate conclusion to be drawn at this point is that dermal or other connective tissue cells of the limb are capable of metaplasia into cartilage, but not muscle, during regeneration; cartilage may be capable of metaplasia into other types of connective tissue. The possibility of metaplastic transformation of muscle remains, but requires further substantiation.

Epidermis

Does the epidermis undergo metaplasia and contribute to the mesenchymal cell population of the blastema? Godlewski (1928) was the first to propose an epidermal contribution to the blastemal mesenchyme, and the idea has since generated a great deal of controversy. In support of epidermal metaplasia, Rose (1948) found that cell numbers decreased in the wound epidermis of regenerating adult newt limbs coincident with an increase in blastemal mesenchymal cells, which could not be accounted for solely on the basis of mitotic activity. He observed tongues of epidermis extending into the blastema, which he interpreted as evidence of metaplasia into mesenchymal cells (Rose, 1948; Rose and Rose, 1965). These epidermal tongues, however, have also been shown to be involved in the phagocytosis of debris, and their presence within the mesenchyme cannot be taken as evidence of a cellular contribution to the latter (Salpeter and Singer, 1960a).

Hay (1952) performed the first cellular marking experiment to test for epidermal metaplasia in regenerating hindlimbs of *Rana pipiens* tadpoles. She homografted fin epidermis from triploid tadpoles to the thighs of skinned diploid hindlimbs, which were then amputated through the metatarsals. The transplanted epidermis migrated distally and formed the covering of the blastema. Hay described 2 of 36 cases in which epidermal cells seemed to transform into fibroblasts in the proximal region of the limb stump, and in one of these cases substantial numbers of triploid cells, presumably from the epidermis, were observed in the blastemal mesenchyme. Rose and Rose (1965) marked the epidermal and blood cells of adult newts (which are among the few adult cell types that divide regularly) by injecting the animals with [^{3}H]thymidine 1 day prior to amputation. The forelimbs were then stripped of skin to a level above the elbow, and amputated just proximal to the wrist. Labeled epidermis migrated down to cover the amputation surface. The limbs regenerated normally, and labeled cells were found in their blastemas among unlabeled ones. The labeled cells were not derived from the blood, as shown by an experiment in which animals were injected with isotope to label epidermal and blood cells, and the labeled skin was replaced with unlabeled skin. After amputation, no labeled cells were found in the resultant blastemas. Rose and Rose concluded that epidermal cells underwent metaplasia into blastemal mesenchyme.

Other investigators, however, have obtained results that lead to the opposite conclusion. O'Steen and Walker (1961) and Hay and Fischman (1961) labeled the epidermis of adult newts with [^{3}H]thymidine, but found no evidence of a contribution of labeled cells to the blastemas and regenerates formed after amputation. In another experiment, Riddiford (1960) amputated adult newt forelimbs bilaterally and then *infused* the resulting blastemas of the left limbs with [^{3}H]thymidine. This technique labeled only the cells of the infused blastema, including the blastemal epidermis. The labeled apical epidermis was then exchanged with the unlabeled apical epidermis of the right blastema, but no labeled cells subsequently appeared in the mesenchyme of the latter.

What is to be made of these conflicting observations and interpretations? The results of Hay's (1952) triploid epidermal transplant experiment could be due to contamination of the grafts by fibroblasts, which in one case contributed to the blastema. The thymidine labeling experiments are more difficult to evaluate. Given that the amount of label administered and the labeling time was adequate to label the nuclei in each of the experiments, the ability to subsequently detect labeled nuclei in the blastema would depend on the specific activity of the label, its dilution by cell division, and the exposure time of the autoradiograms. However, the values for these parameters were quite different in the various experiments. Rose and Rose (1965) used [^{3}H]thymidine with a specific activity of 1.8 Ci/mmole and exposed their slides for 3–12 weeks. The specific activities and exposure times for the other investigators were 0.36–3.0 Ci/mmole and 3 weeks (Hay and Fischman, 1961), specific activity not reported and 3–12 weeks (O'Steen and Walker, 1961), and 0.89 Ci/mmole and 3–5 weeks (Riddiford, 1960). O'Steen and Walker (1961) and Rose and Rose (1965) thus used the same exposure times, but direct comparison of their results cannot be made, because the former did not report the specific activity. Rose and Rose (1965) believe that Hay and Fischman (1961) could not detect an epidermal contribution to the blastema, because their labeling was not heavy enough to compensate for dilution due to cell division. The specific activity of Hay and Fischman's (1961) [^{3}H]thymidine was, at least in some cases, 1.7 times greater than Rose and Rose's, but the exposure times for their autoradiograms were much less. Rose and Rose's own cases show that the label is extensively diluted during blastema formation, and the mesenchyme cells that they believe to be derived from the epidermis have only four to eight grains per nucleus. They state, however, that this number is easily distinguished from the very low background in their experiment. The specific activity in Riddiford's (1960) experiment was half that of Rose and Rose's, and the exposure time much shorter. But dilution was not as much of a factor in her experiment, because labeled epidermis was immediately available for incorporation into the blastemal mesenchyme without an intervening period of cell division. Furthermore, Riddiford's photographs indicate that the grain density over the grafted apical epidermis in her experiments was very high. However, Rose and Rose (1965) point out that the labeled apical epidermis may not have been able to contribute to the blastema because these cells would be displaced outward and sloughed owing to continual distal migration of more proximal, unlabeled epidermal cells.

My conclusion is that none of these experiments conclusively demonstrate whether the epidermis does or does not contribute to the blastemal mesenchyme. More than anything else, they underline the difficulties of using [^{3}H]thymidine as a marker for epidermal cells that undergo extensive division and are lost by migration. However, Namenwirth (1974) has also used polyploidy as a cell marker to test the potency of pure epidermal cells during regeneration. She removed the epidermis from the limb skin of triploid axolotls by trypsin and collagenase treatment. This epidermis was grafted in place of the skin of x-irradiated limbs of diploid animals and the limbs amputated through the grafted regions. These limbs did not regenerate, and the wound epidermis that covered the amputation surface was confirmed to be triploid. This result clearly shows that unirradiated epidermis alone cannot form a blastemal mesenchyme on an irradiated limb stump, but does not rule out the possibility that epidermis might normally contribute some cells to the blastema. A similar experiment should be done grafting polyploid pure epidermis to unirradiated diploid limbs to determine whether triploid cells might contribute to the blastema under normal conditions.

Can Unirradiated Skin Reverse Irradiation Damage?

Irradiation of a larval or adult salamander limb with appropriate doses of x-rays destroys the ability of the limb to regenerate by causing chromosomal damage, thus interfering with normal cell division and decreasing cell survival (see Wallace, 1981, for review). Regeneration can take place, however, if unirradiated tissues are grafted into an irradiated limb. Stinson (1964) grafted a half zeugopodium (lower arm) into the irradiated stylopodium (upper arm) of adult newt forelimbs and obtained regenerates that were half zeugopodia. When skin, cartilage, muscle, or brachial nerve was grafted from the limbs of white axolotls to the irradiated limbs of dark axolotls, or vice versa, the regenerates formed after amputation through the graft region were always of donor color (Umanski, 1938; Maden and Wallace, 1975, 1976). Dunis and Namenwirth (1977) grafted triploid skin to irradiated diploid limbs of axolotls and showed that a large fraction of the regenerate cells were derived from the graft. These grafts were also done between forelimb and hindlimb, and there was a high correlation between the morphology of the regenerate and the source of the graft. All these results suggest that grafted tissues, including the skin, enable irradiated limbs to regenerate because they act as a source of healthy cells from which to form a blastema.

However, in Dunis and Namenwirth's (1977) experiment, the frequency of triploid cells in the regenerates was less than expected when compared to control triploid limbs, leading the authors to conclude that not all the cells in the regenerate may have come from the graft. The animals had been irradiated across the body, so there was little likelihood that distal migration of diploid cells from healthy proximal host tissue could account for the result. The only other source of diploid cells would be irradiated host limb tissue, suggesting that graft cells somehow rescued irradiated host cells from radiation damage so that they could participate in regeneration. This idea has received apparent support from grafting experiments in which radioactive labels have been employed to distinguish between graft and host cells.

Desselle (1968) grafted [^{3}H]thymidine or ^{32}P-labeled limb cartilage into x-rayed limbs of adult newts and detected unlabeled (presumably host) cells among the labeled cells of the blastemas formed after amputation through the graft region. The reverse experiment was done by Desha (1974). Regenerating adult newt forelimbs were first labeled with [^{3}H]thymidine as they were redifferentiating; the labeled regenerates were then x-rayed and provided with grafts of unirradiated skin. Several of these limbs regenerated after reamputation, and labeled host cells were found in the proximal part of the blastema. Rose and Rose (1965) labeled the epidermis of adult newts 1 day prior to x-irradiating their lower arms and then stripped the skin from the limbs to a level above the elbow and amputated them just proximal to the wrist. Unirradiated, labeled epidermis migrated after the irradiated area. Regeneration took place in some of these limbs, and some lightly labeled (the initial label became heavily diluted) cells were found among unlabeled ones in the blastema. Their interpretation was that not only did the epidermis contribute to the regeneration blastema, but induced irradiated internal tissues to do so.

It is perhaps not surprising that some irradiated cells might become part of a blastema derived mainly from implanted nonirradiated tissue. There is ample histological evidence that irradiated limb tissues undergo normal cellular dedifferentiation after amputation (Butler, 1933; Lazard, 1968; Desha, 1974; Maden and Wallace, 1976; Wertz and Donaldson, 1980) and that the dedifferentiated cells even synthesize DNA (Wertz and Donaldson, 1980). But the amount of DNA synthesis and mitotic activity in amputated irradiated limbs is severely depressed compared to controls, and there is clear evidence of radiation damage in the form of vacuolated cells, micronuclei, and cell loss (Maden and Wallace, 1976; Wertz and Donaldson, 1980). It is possible that irradiated, dedifferentiated cells that survive might be incorporated into a regenerate formed mainly from healthy graft tissues, but not proliferate. Such a limited host contribution could explain why fewer triploid cells than expected were found in the regenerates obtained from skin-grafted irradiated limbs by Dunis and Namenwirth (1977).

The real question is whether graft cells can effect repair of x-ray-induced host cell chromosomal damage so that they can proliferate to form a large part of most of a regenerate. There are some experimental results that can be interpreted in this way. For example, Trampusch (1959) grafted limb skin from dark axolotls to the irradiated limbs of white axolotls. Most of the 104 regenerates obtained were of donor origin (dark), but 3 were white, suggesting a substantial host origin of the blastema in these cases. Clarke (1978) grafted forelimb skin to x-rayed hindlimbs of adult newts. After amputation, half the cases regenerated as hindlimbs, again suggesting that host cells made a major contribution to the regenerates. Desselle and Gontcharoff (1978) implanted limb cartilage from *D. fuscus* into the irradiated limbs of *T. cristatus* and amputated through the graft. The nuclei of *T. cristatus* have a bit more than twice the DNA content of *D. fuscus* nuclei, and cytophotometric analysis was used to distinguish nuclei of the two species in the regenerates. Misidentification of postreplication donor cells as G_1-blocked host cells was ruled out in a separate experiment. A total of 95% of the regenerate cells were identified as host, and the absorption measurements indicated the presence of nuclei with 4C DNA content. Surprisingly, the absorption values of the donor nuclei exhibited little variation from the 2C value, suggesting that donor cells divided very infrequently or not at all.

However, none of these experimental results are conclusive. With the exception of Trampusch's (1959) experiment, only the limbs were irradiated, leaving the girdle shielded, as has been the case in most of the experiments suggesting recovery from irradiation damage. X-irradiation of limbs causes temporary tissue damage and regressive changes even in the absence of amputation (Wertz and Donaldson, 1980). The limbs in most of the irradiation recovery experiments were amputated before these changes subsided (a matter of several weeks). Hence it is possible that dedifferentiative activity may have spread into the girdle region, mobilizing cells that migrated distally to divide and make up the majority of the regeneration blastema. Furthermore, grafting normal tissues to an irradiated host limb necessitates making a large wound in either the host or donor tissue. Wallace et al. (1971) have shown that the simple act of making a longitudinal surface wound in the irradiated and amputated limbs of axolotls allows some limbs to regenerate, probably by distal migration of unirradiated proximal cells along the wound.

The results of other experiments fail to support the notion of mitosis of irradiated cells in the presence of unirradiated cells. The cells of irradiated axolotl limb blastemas appear to be blocked in the G_1 phase of the cell cycle (Maden, 1979b). Irradiated diploid limbs of axolotls that were also irradiated across the body failed to regenerate when supplied with a graft of unirradiated triploid epidermis, which was verified to have not been replaced by irradiated host epidermis (Namenwirth, 1974). Blastemal extracts infused into irradiated and amputated adult newt limbs fail to restore regenerative capacity, although they do promote the formation of pseudoblastemas (Deck and Dent, 1970). Regenerates arising from irradiated tails provided with unirradiated limb skin, or vice versa, always have the organ morphology of the donor graft (Trampusch, 1959; Hofmann et al., 1978). Finally, Maden and Wallace (1976) have pointed out that if rescue of irradiated cells by unirradiated ones is possible, irradiation damage to local limb regions should be reversible by adjacent healthy tissues. This is not the case; regeneration fails to occur after amputation through a locally irradiated limb region, and a normal regeneration blastema grafted to a more proximal level of an irradiated limb stump fails to evoke intercalary regeneration of the intermediate structures from the host stump, as it normally does (Maden, 1980).

In summary, there seems to be little evidence that grafts of skin or other tissues can stimulate the recovery of limb cells from x-ray-induced damage so that they may divide and make a substantial contribution to a regenerate.

Pattern Information Is Contained in the Skin

One of the most fascinating and central problems of regeneration is to understand the morphogenetic mechanism by which missing limb parts are recreated out of the growing blastema. It was once believed that the blastema is a pluripotent cell mass capable of forming a variety of non-limb structures, and that the specific structure and asymmetry it redifferentiates is dictated by specific inductive signals emanating from adjacent differentiated tissues (for reviews see Goss, 1961; Stocum, 1978, 1984; Wallace, 1981). It is now apparent that enough morphogenetic information to specify the structural pattern and handedness of the limb regenerate is inherited by blastema cells directly from their differentiated parent cells and can be expressed independently of adjacent stump cells (Stocum, 1968a,b, 1978, 1980a, 1982, 1984; Stocum and Dearlove, 1972; Stocum and Melton, 1977; Polezhaev, 1979; Wallace, 1981). This information is contained in the skin (as well as in other tissues of the limb) and is asymmetrically distributed around its circumference and along its proximodistal axis.

The Skin Contains a Complete Set of Pattern-Regulating Information

Removal of the anterior or posterior half of an adult newt or larval salamander zeugopodium or suppression by x-irradiation of the ability of these halves to regenerate results in the regeneration of a half limb from the remaining or nonsuppressed half after amputation (Goss, 1957a,b; Maden, 1979b). However, if just the bone and muscle of one of the halves is removed, leaving the skin intact, a complete limb is regenerated from the remaining half after amputation (Weiss, 1926; Goss, 1957a). Dunis and Namenwirth (1977) have shown that a cuff of skin by itself can give rise to a regenerate of normal form with skeleton. Collectively, these results demonstrate that a complete circumference of skin contains, at any proximodistal level of amputation, all the pattern-regulating information required to make a regenerate of normal form and asymmetry, excluding the formation of muscle (Dunis and Namenwirth, 1977).

Pattern-Regulating Information of the Skin Is Organ Specific

Heterotopic grafting experiments have demonstrated that the morphogenetic information of the skin is organ specific. If tail skin is grafted onto irradiated limbs, or limb skin onto irradiated tails, the regenerates formed after amputation are entirely of graft character (Trampusch, 1958a,b). However, if tail skin is grafted over normal internal limb tissues, tail–limb chimeras are obtained after amputation through the graft (Glade, 1963). The tail and limb portions of the chimeras are derived from the corresponding graft and host tissues. Thus, alone or in combination, tail and limb skin produce regenerates with a morphology corresponding to the organs of origin.

Pattern-Regulating Information Is Asymmetrically Distributed in the Skin

Experiments on axolotls and adult newts, in which cuffs or sectors of limb skin have been displaced 180° relative to the internal tissues, indicate, in accord with the anatomical asymmetry of the skin, that its morphogenetic information is asymmetrically distributed in the transverse axes and the proximodistal axis. Circumferential 180° translocations of skin juxtapose the skin and internal tissues of opposite poles of the anteroposterior and/or dorsoventral axis. After amputation, supernumerary limbs, of various degrees of completeness, arise at the sites of polar confrontation (Droin, 1959; Rahmani, 1960; Carlson, 1974b, 1975; Settles, 1978; Slack, 1980a, 1983). Supernumeraries are also formed after cross-transplanting muscles (Carlson, 1975). Simple removal and replacement of skin or muscles in their original locations does not elicit supernumeraries. Supernumerary formation must therefore be explained by assuming the existence of different morphogenetic qualities at opposite axial poles of the skin and internal tissues. When polar opposite qualities are apposed by grafting skin onto internal tissue of the opposite side of the limb, they interact in some way to produce accessory morphogenetic fields (see Carlson, 1975; French et al., 1976). Lheureux (1975b) demonstrated that interactions between cells carrying different circumferential morphogenetic qualities are absolutely essential for regeneration after simple amputation. He grafted longitudinal strips of skin onto irradiated limbs so that just one skin anatomical quality (dorsal or ventral) was present at the amputation surface or two opposite qualities (dorsal *and* ventral) were present. Regeneration took place when two qualities were present, but not when one was present.

The strength of the interactions between skin and internal tissues of opposite axial poles is nonequivalent. This fact is most clearly seen in experiments on the zeugopodium (lower arm or leg), in which anterior, posterior, dorsal, or ventral skin sectors have been grafted onto the polar opposite internal tissues. As calculated from the combined data of Lheureux (1972) and Maden and Mustafa (1982), the effectiveness of these four polar combinations in eliciting supernumerary formation (expressed as the percentage of cases that formed supernumeraries in each combination) decreased in the following order: posterior skin on anterior internal tissue (63%), anterior skin on posterior internal tissue (24%), dorsal skin on ventral internal tissue (20%), and ventral skin on dorsal internal tissue (5%). Thus posterior skin on anterior internal tissue is by far the most potent combination, and ventral skin on dorsal internal tissue the least. It is also clear that there is a directionality to the strength of the interaction within each axis, posterior to anterior and dorsal to ventral being more effective than anterior to posterior and ventral to dorsal.

Data obtained by making the same grafts in the stylopodium (upper arm or leg) (Tank, 1979; Maden and Mustafa, 1982; Rollman-Dinsmore and Bryant, 1982) are not as clear. The main complication is that different investigators have used different anatomical landmarks to designate axes in the stylopodium, whereas everyone has used the same landmarks to designate axes in the zeugopodium (see Fox, 1982). Rollman-Dinsmore and Bryant (1982) designate the line bisecting the extensor and flexor muscles as the dorsoventral axis, and the line perpendicular to that as the anteroposterior axis. Tank (1979) and Maden and Mustafa (1982) reverse these axial definitions. Hence the data on frequency of supernumerary formation have been interpreted differently, depending on which axial designations have been used. If the results of all these investigators are combined and expressed in terms of the Tank and Maden–Mustafa axial definitions, the combinations of dorsal skin on ventral internal tissue and posterior skin on anterior internal tissue are equivalent in strength (34 and 35% supernumerary formation, respectively), followed by ventral skin on dorsal internal tissue (23%) and anterior skin on posterior internal tissue (14%). This order is quite different from that of the zeugopodium. However, if the stylopodial data are expressed in terms of the axial definitions of Rollman-Dinsmore and Bryant, the order more closely approximates that of the zeugopodium: posterior skin on anterior internal tissue (37%), dorsal skin on ventral internal tissue (28%), anterior skin on posterior internal tissue (21%), ventral skin on dorsal internal tissue (19%). Nevertheless, in both cases the posterior to anterior and dorsal to ventral directionality of effect within the anteroposterior and dorsoventral axes is observed.

What these data mean in the context of various theories of pattern regulation is at present unclear. One interpretation is that the posterior and dorsal skin represent the high points in two hierarchies of biochemical switches that specify pattern asymmetry across the anteroposterior and dorsoventral axes, as proposed by Slack (1980a,b, 1983). Further studies are needed, however, before this interpretation can be accepted. One problem is that if the data of Rollman-Dinsmore and Bryant (1982) and those of Tank (1979) and Maden and Mustafa (1982) are arranged so that axial definitions are equivalent, but the three studies considered separately, the hierarchy of frequency of supernumerary formation is completely different in each, and the intraxial directionality of effect is different or nonexistent. What is called for is a single large study involving skin and muscle grafts in both the stylopodium and zeugopodium, in which the axial definitions of both segments are made equivalent by basing them on the position of the extensor and flexor muscle groups when the limb is straightened and supinated (the "anatomical position").

The presence of a proximal to distal sequence of morphogenetic information in the limb skin has been demonstrated by grafting sleeves of skin to more proximal or distal levels in normal or irradiated forelimbs of adult newts and amputating through the graft (Lheureux, 1975a,b). The zeugopodial bones (radius and ulna) are shorter than the stylopodial bone (humerus). In normal forelimbs, zeugopodial skin grafted to the stylopodium tends to decrease the length ratio of humerus/ulna in the resulting regenerates with respect to the control ratio, whereas stylopodial skin grafted to the zeugopodium increases the ratio. Amputation through an irradiated zeugopodium that has received a graft of unirradiated stylopodial skin results in longer than normal limbs with serial repetitions of humerus and radius–ulna; the reverse combination results in shortened limbs missing the distal part of the humerus and the proximal part of the zeugopodial bones. This result shows clearly that different proximal–distal levels of the skin carry nonequivalent morphogenetic information that determines that only those structures distal to the level of amputation can be regenerated.

The Dermis Carries the Asymmetric Information

A number of experiments have shown that it is the dermis, and not the epidermis, that carries the asymmetric morphogenetic qualities of the skin. Glade (1963) separated tail epidermis and dermis and grafted them separately to skinned limbs. Limbs receiving tail epidermis always regenerated normal limbs after amputation, whereas those receiving tail dermis regenerated tail–limb chimeras, even though the tail dermis became covered by limb wound epidermis. Carlson (1975) removed the epidermis from axolotl forelimb skin and then grafted the dermal cuff back on the limb after 180° rotation around its longitudinal axis. Multiple regenerates resulted after amputation through the graft in 82% of the cases. To show that epidermal displacement does not evoke supernumerary formation, Carlson (1975) performed a two-step experiment. A long sleeve of skin was rotated 180°. Then a short skin sleeve was removed from the middle of the original sleeve, and the wound allowed to heal for 37 days. New dermis of normal asymmetry (presumably formed by local fibroblasts) was thus overlain by rotated epidermis, but 94% of the resulting regenerates were single limbs. Tank (1981) displaced fragments of whole or minced dermis 180° and obtained a high frequency of supernumerary formation after amputation through the grafts. No supernumeraries formed, however, if the displaced dermis had first been x-irradiated or repeatedly frozen and thawed. Provided that these treatments do not alter the normal function of the matrix, this result suggests that living cells, not the matrix, of the dermis are required for the pattern-regulating interactions of supernumerary formation. It also predicts that displacements of limb skin in young larvae would not affect the normal course of regeneration, since the dermis of that skin is acellular. This prediction agrees with Holtzer's (1958) observation that limb and tail skin of larval salamanders exerted no morphogenetic influence on regeneration after their exchange.

Role of the Epidermis in Limb Regeneration

Although there is no evidence that the epidermis contributes to the mesenchyme of the blastema, there is extensive evidence that its role in regeneration is much more than that of a passive cover protecting the blastema from the external environment. Singer and Salpeter (1961) have described a series of observations and experiments indicating that the epidermis functions phagocytically to remove debris created by cell death, or foreign tissue and inorganic matter inserted into the blastema. Beyond this, however, contact between the epidermis and underlying mesenchymal tissues is essential for blastema formation, outgrowth, and possibly pattern specification, as discussed below.

The Epidermis Promotes Blastema Formation and Outgrowth

Interaction between the wound epidermis and the dedifferentiated cells of subjacent mesodermal tissues is essential for blastema formation. This fact is clearly demonstrated by experiments in which contact between epidermis and mesodermal cells is prevented. If freshly amputated limb or tail wound surfaces are covered by full-thickness skin, a blastema fails to form and regeneration is inhibited, even though dedifferentiation takes place (Tornier, 1906; Taube, 1921; Godlewski, 1928; Polezhaev and Faworina, 1935; Mescher, 1976; Tassava and Garling, 1979). Likewise, blastema formation is prevented by insertion of freshly amputated limb tips into the coelom, the environment of which inhibits epidermal wound closure (Goss, 1956a; Pietrzyk-Walknowska, 1959; Loyd and Tassava, 1980), but if a limb tip is inserted into the coelom after epidermal closure has occurred, regeneration proceeds normally. Blastema formation is also prevented if the wound epidermis is irradiated with ultraviolet light (Thornton, 1958).

Epidermal–mesodermal interaction is also essential for the continued growth of the blastema, once it has formed, and hence the proximal to distal order of appearance of regenerated parts. This has been shown by removing the epidermis from progressively older blastemas and then sewing the limbs into a skin pocket or the coelom (Polezhaev and Faworina, 1935; Goss, 1956b) and by implanting epidermis-free blastemas into a tunnel bored in the connective tissue of the dorsal fin (Stocum and Dearlove, 1972). The more advanced the stage of the blastema at the time of the operation, the more complete is its ultimate morphogenesis along the proximodistal axis. Blastemas of any stage implanted into a fin tunnel so that their distal tips protrude and are recovered with fin epidermis develop a complete proximodistal array of limb parts (Stocum and Dearlove, 1972).

The effect of the epidermis on underlying blastema cells is presumably mediated by secretion of a product or products that diffuse proximally, maintaining a zone of undifferentiated cells as the blastema grows and redifferentiates new limb parts in a proximal to distal direction (Smith et al., 1974; Tassava and Mescher, 1975; Globus et al., 1980). Singer and Salpeter (1961) noted that the basal cells of the wound epidermis of regenerating adult newt limbs exhibited ultrastructural evidence of secretory activity by virtue of their abundant rough endoplasmic reticulum, and they also observed a fuzzy cortical zone in these cells possibly associated with secretion. Wound epidermis is highly active in RNA and protein synthesis (Bodemer, 1963; Bodemer and Everett, 1959; Anton, 1965; Hay, 1966a), and blastema formation is greatly retarded in limbs whose wound epidermis is derived from skin treated with actinomycin D (Carlson, 1969).

Other biochemical evidence strongly supports the notion that the epidermis is involved in the production of substances associated with the dedifferentiated state. Synthesis of the proteolytic enzyme collagenase and the glycosaminoglycan hyaluronic acid is increased during blastema formation and growth, whereas the synthesis of collagen and hyaluronidase is decreased; during redifferentiation this synthetic pattern is reversed (Grillo et al., 1968; Dresden and Gross, 1970; Toole and Gross, 1971; Johnson and Schmidt, 1974; Linsenmayer and Smith, 1976; Mailman and Dresden, 1976). Collagenase activity is highest at the distal end of the blastema, closest to the apical wound epidermis that is not underlain by a basement membrane (Grillo et al., 1968). The cell type producing collagenase has not been identified in the limb, but in the frog tail undergoing metamorphosis it is the epidermis (Eisen and Gross, 1965). The cell type producing hyaluronate and hyaluronidase in the regenerating limb is also unknown.

Chapron (1974) has presented some evidence, based on [^{3}H]fucose labeling, that glycoprotein in regenerating adult newt limbs is synthesized first in the epidermis and then is secreted into the extracellular spaces of the blastema. This labeled glycoprotein was not identified, but may consist largely of fibronectin, which is present in the basal layer of the wound epidermis and in the underlying blastema, but not in the epidermis of proximal limb skin (Repesh and Furcht, 1982; Gulati et al., 1983). Hence collagenase, hyaluronate, and fibronectin may all be secretory products of the wound epidermis (or blastema cells may be influenced by the epidermis to produce them) that in combination provide an environment conducive to maintaining blastema cells in an undifferentiated and dividing state so that blastemal outgrowth can occur.

The Epidermis Imposes Directionality on Blastemal Outgrowth

The function of the epidermis in blastema formation and outgrowth is not restricted to just an effect on blastema cell differentiation. The apical cap of the wound epidermis gives a directionality to the outgrowth of the blastema. If a piece of skin is excised from one side of the tip of a larval limb in the dedifferentiative stage of regeneration, the apical cap is repositioned at an angle to the longitudinal axis of the limb by the migration of the epidermis closing the wound. The blastema cells then accumulate under the cap and the resulting regenerate redifferentiates at an angle to the limb stump (Thornton, 1960, 1962). Likewise, an apical cap autografted from the tip of a young blastema to its base causes the lateral outgrowth of a supernumerary limb at the transplant site (Thornton and Thornton, 1965).

The Epidermis May Also Have a Role in Pattern Specification

The necessity of the epidermis for morphogenesis of the blastema is usually attributed to its outgrowth-promoting function, with actual pattern specification being a function of the asymmetrical morphogenetic information carried by mesodermal cells. This follows from the experimental demonstration, discussed earlier, that no asymmetry of morphogenetic quality can be detected in the axes of the epidermis (Carlson, 1975). Nevertheless, two facts suggest that the epidermis might play a role in specification of regenerate pattern: First, mitosis of blastema cells is dependent on innervation (see Chap. 5 for further discussion), so that if a larval or adult salamander limb is totally or partially denervated after a blastema has formed, growth in blastemal volume is abolished or reduced. In spite of this, the blastema can often elongate and undergo size-invariant regulation to produce a correspondingly smaller regenerate (Singer and Craven, 1948; Powell, 1969; Maden, 1981). The size reduction of skeletal elements follows a proximal to distal gradient, being greatest in the stylopodium, less in the zeugopodium, and least in the autopodium (Maden, 1981). Second, epidermis-free blastemas implanted into dorsal fin tunnels also form regenerates that are subnormal in size, but these regenerates are *distally truncated*, whereas regenerates derived from implanted blastemas allowed to become recovered with fin epidermis are distally complete (Stocum and Dearlove, 1972). The fact that abolished or reduced blastemal growth in the presence of epidermis (denervated limbs) allows the development of smaller but distally complete regenerates, whereas in the absence of epidermis it results in smaller but distally truncated regenerates, strongly suggests that the epidermis functions as a boundary telling the mesenchymal cells where the tip of the regenerate is to be (Maden, 1977; Stocum, 1978, 1980b). The actual size reductions of the skeletal elements in the truncated and nontruncated cases, however, have not been measured. Such measurements will be necessary to determine the validity of this idea, which predicts that there would be the same number of cells in truncated and nontruncated regenerates, but the skeletal elements of the latter would be smaller.

A Morphogenetic Model for Limb Regeneration

The foregoing discussion suggests a three-dimensional topological model for limb regeneration (see Stocum, 1983) that emphasizes the importance of skin components. This model, the general outlines of which are illustrated in Figure 3.1, incorporates elements of the two-dimensional polar coordinate model of Bryant et al. (1981) and the averaging model of Maden (1977). Cells at a given proximodistal level are assumed to have three positional values (Wolpert, 1969, 1971) encoded on their surface that tell them their location with respect to other cells in three-dimensional space. Two of these values specify position in the transverse plane; the third specifies position along the proximodistal axis. The positions of dedifferentiated cells relative to cells carrying boundary values that outline the regenerate determines their pattern of redifferentiation. For simplicity, a polar coordinate system is used to map positional values (radial and angular) in the transverse plane, with the angular values represented as clock numbers, as in the polar coordinate model (French et al., 1976). The proximodistal coordinate is orthogonal; however, any other coordinate system that reflects the experimental data would work as well.

The model is based on four ideas: (1) The dermis of the skin, muscle, and cartilage all carry an asymmetrical pattern of radial, circumferential, and proximodistal positional values that is inherited by the blastema cells derived from them; (2) blastema cells remember their proximodistal value of origin and normally can change that value only in a distal direction (distal transformation) [this memory constitutes the proximal boundary of what is to be regenerated (Pescitelli and Stocum, 1980)]; (3) the various mesodermal cell types can undergo metaplastic transformation among themselves; and (4) the wound epidermis represents the periphery of the regenerate (i.e., the most radial and most distal positional values) everywhere on the limb. Hence the morphogenetic information of the wound epidermis is not asymmetrical, but consists only of a single quality representing the *external boundary* of what is to be regenerated (Maden, 1977).

This organization defines a *morphogenetic field*, whose classic property is its ability to make adjustments (regulate) in its remaining parts to eliminate discontinuities created by loss of some of its positional values. In the case of the regenerate field described here, elimination of the discontinuity is accomplished epimorphically by *intercalary regeneration* (Iten and Bryant, 1975; Stocum, 1975b), which consists of cell division at the edges of the discontinuity and the assignment of missing positional values to progeny cells by intercellular communication, using the proximal and external boundaries as ultimate reference points.

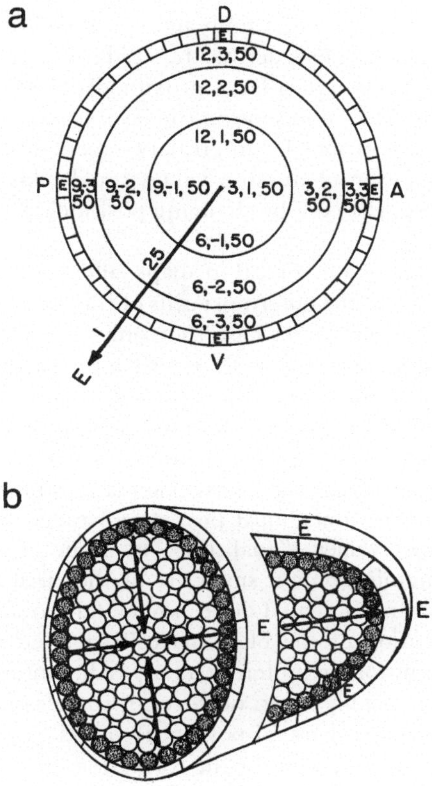

FIG. 3.1. Three-dimensional boundary model of regenerate morphogenesis. (a) Diagrammatic cross section through the upper arm or leg. The inner circle represents the stylopodial cartilage, the next circle represents muscle, and the outer circle represents dermis and subcutaneous connective tissue. These tissues are covered externally by the epidermis (E). The epidermis has a single positional value representing the external boundary of the regenerate. The cells of the mesodermal tissues each carry three positional values, represented by the numbers. The first number in each trio is the angular position, the second number is the radial position, and the third number is the proximal–distal position. The arrow represents the proximal–distal axis (D, dorsal; V, ventral; A, anterior; P, posterior). During normal regeneration, missing proximal–distal values are intercalated between the level of origin (50) and the epidermis (E). Radial and angular intercalation also occurs at the tip of the blastema (see text), but for simplicity is not represented here. (b) Cutaway diagram of a regeneration blastema derived from skin grafted to an irradiated limb, illustrating how the blastema regulates to restore missing positional values in the cross-sectional plane. The blastema cells are derived from the dermis and represent all of the angular values at the base of the blastema, but only the most peripheral radial values (stippled cells). Their progeny (unstippled cells) intercalate the missing radial values to restore the cross-sectional pattern (indicated by the set of four arrows), while the missing proximal–distal values are simultaneously intercalated between the level of origin and the epidermis (indicated by the remaining arrow) (for further details see Stocum 1980b, 1983).

After simple amputation of a limb and initial dedifferentiation, there is no discontinuity in radial values across the epidermis and the underlying blastemal mesenchyme. There is a large proximodistal discontinuity between the epidermis and the value representing the level of origin of the blastema cells, and this will be eliminated by intercalary regeneration between these two boundaries. Restoration of missing values could take place by an averaging mechanism (Maden, 1977) or by changing positional value distally, one unit at a time.

The growing blastema is conical in shape, and, because of this geometry, all of the angular values of the outermost cells of the blastema (which will become dermis) will be present at its base, but only a few angular values will be represented at its tip, owing to space restrictions. Furthermore, some internal radial values may be missing at the tip for the same reason. Hence, as suggested by Bryant et al. (1981), angular (and radial) intercalation would have to occur at the apex of the blastema in order to maintain a complete map of positional values specifying the transverse plane. This idea is consistent with the fact that irradiated limbs must possess unirradiated skin with sufficiently disparate positional values ("qualities") in order to regenerate (Lheureux, 1975b). Skin containing only a small arc of the circumferential sequence of values grafted to an irradiated limb would fail to regenerate because the positional information of the blastema cells derived from it would not be sufficiently different for circumferential and radial intercalation to occur. Such intercalation would readily occur, however, if the blastema were derived from skin containing widely separated positional values.

This model can account well for the formation of a normal regenerate by a limb from which internal tissues such as cartilage or muscle have been removed. As long as the external boundary (epidermis) of the regenerate field is present, intercalary regeneration of any missing transverse, as well as proximal–distal, positional values can take place (see Stocum, 1983). Thus the model emphasizes the importance of the epidermis for regenerate morphogenesis, in addition to its outgrowth-promoting role.

Conclusions

Amphibian limb regeneration is highly dependent on the skin. The epidermis seals the amputation surface and interacts with underlying dedifferentiated mesodermal cells to promote outgrowth of a regeneration blastema. The epidermis might also contain a single morphogenetic quality (positional value) that functions as the external boundary of the regeneration field. There is no evidence, however, that the epidermis contributes dedifferentiated cells to the blastemal mesenchyme.

The dermis of the skin (along with the internal mesodermal tissues) does contribute dedifferentiated cells to the blastemal mesenchyme and contains an asymmetrical pattern of morphogenetic qualities that reflect the anatomical asymmetries of the anteroposterior, dorsoventral, and proximodistal axes. Interaction between at least two different qualities is an essential requirement for regeneration to take place.

One model to account for regeneration, which stresses the role of the epidermis in pattern specifications, postulates that pattern regulation takes place in three dimensions via local interactions between cells carrying different positional values. Missing positional values are reassigned to mesenchyme cells by reference to a three-dimensional set of boundary values representing the outline of what is to be regenerated.

A number of important questions about the role of the skin in regeneration remain unanswered:

1. What is the significance, to the mechanism of pattern regulation, of the differences in frequency of supernumerary formation elictied by different combinations of polar opposites of skin and internal tissues?
2. Does the skin contain pattern information at all stages of its development? Is there any correlation between the ability of grafted skin to elicit supernumerary formation and the developmental stage of the skin?
3. What is the cellular and biochemical basis of the epidermal–mesenchymal interaction that maintains mesenchymal cells in a dedifferentiated state so that blastemal outgrowth and pattern specification can take place?

Answers to these questions will do much to further our understanding of how the skin participates in regeneration.

4 The Hormonal Milieu in Amphibian Appendage Regeneration

RICHARD A. LIVERSAGE, DANIELLE S. MCLAUGHLIN,
AND HOOLEY M. G. MCLAUGHLIN

During the normal course of adult newt appendage regeneration, epidermal wound healing of the amputation surface ensues and an apical epidermal cap forms. This occurs as the cellular debris from the injury within the appendage is phagocytized during the first few days postamputation. From the 5th to the 20th day after amputation there is considerable loss of structure (dedifferentiation) within the distalmost tissues of the amputated forelimb (Schotté and Hall, 1952; see also Butler, 1933, 1935; Thornton, 1938; Butler and Schotté, 1941). Cells of the differentiated tissues become freed from their intercellular matrices, lose integral qualities, and acquire embryonic-like characteristics: prominent nucleoli, an increased nuclear–cytoplasmic ratio, and an increase in free ribosomes. These seed cells are the source of the regeneration blastema cells. Almost immediately these cells re-enter the cell cycle, under the influence of nerves, and proliferate (Mescher and Tassava, 1975; Globus and Liversage, 1975a,b; Globus, 1978) to form the "accumulation blastema" (Butler and Puckett, 1940) at the tip of the amputated appendage. Further consideration of the normal stages of forelimb regeneration in adult *Notophthalmus viridescens* can be found in the descriptions and illustrations of Schotté and Hall (1952), Schotté and Liversage (1959), Liversage and Scadding (1969), and Iten and Bryant (1973).

A "multiple hormonal background" exists from which we have derived the more unified concept of the *Hormonal Milieu* that is involved in growth and differentiation in adult newt appendage regeneration. The new structure and environment, established during the early stages of regeneration, is the framework upon which the Hormonal Milieu acts. The early experiments provide the basis for our approach to the study of hormones in regeneration. The environment in which hormones operate, in part created by the hormonal action itself, determines the activity of the Hormonal Milieu. The term *Hormonal Milieu* distinguishes the hormonal background as an integral entity within the organism; hormonal systems interact with one another and their surrounding local (micro-)environments to form a larger interrelated system, the Milieu.

One of the first studies of hormones in urodele appendage regeneration was initiated by Walter (1910). He observed that stumped or stunted, malformed regenerates resulted 60 days after concomitant amputation and bilateral thyroidectomy in adult *Triton cristatus*. In 1926 Schotté first demonstrated the essential role of the adenohypophysis in adult *T. cristatus* forelimb regeneration.

In the amputation site of the adult newt appendage following hypophysectomy, the typical regenerate structure and environment are not present. The rate and degree of epidermal wound healing of the amputation site occur normally, and what appears to be an apical cap forms. But only limited tissue dedifferentiation (Hall and Schotté, 1951; Liversage and Scadding, 1969, Fig. 7a–c) of the cut mesodermal stump tissues occurs, and a fibrocellular pad, or cicatrix, forms as a result of the cellular activity. Also, the dermis encroaches precociously, as evidenced by the presence of dermal skin glands differentiating over the wound surface (Hall and Schotté, 1951; Liversage and Scadding, 1969; Liversage and Fisher, 1974). Dedifferentiation also involves the more proximal tissues of the stump (about 2 mm from the amputation site), uninjured during amputation. Following limb or tail amputation in previously hypophysectomized adult newts, dedifferentiation of the soft and hard tissues is not as extensive as in controls (Hall and Schotté, 1951; Vethamany-Globus and Liversage, 1973b), presumably owing to systemic hormonal imbalance (Liversage and Price, 1973). It is noteworthy that a fibrocellular cicatrix forms following amputation of a forelimb, hindlimb, or tail in hypophysectomized, thyroidectomized, and singly hypophysectomized or thyroidectomized adult newts (Richardson, 1940, 1945; Hall and Schotté, 1951; Liversage and Scadding, 1969; Vethamany-Globus and Liversage, 1973b), suggesting that some limited dedifferentiation has occurred. The cicatrix and dermis tend to seal the wound, preventing cell–cell contact between the epidermis and the cut mesodermal stump tissues.

Factors in Regeneration

There are various physiological factors in adult newt appendage regeneration that are essential for the new development. Amputation injury, epidermal contact, hormones, and nerves are all elements that play a role in the visible manifestation of regenerative activity. These factors act as a "series of epiphenomena" (O. E. Schotté, 1951, personal communication), in concert with intracellular components of dedifferentiated cells, in epimorphic regeneration (Carlson, 1970, 1978). The genetic and basic cytoplasmic components of the cells of an appendage in an adult animal remain following ontogenetic development. Following injury (amputation), the influence exerted upon a cell by its microenvironment, which includes the Hormonal Milieu, is a major influencing and determining factor or "key" as to whether regeneration will ensue.

Systemic and Local Stress Due to Injury

The endocrine glands of vertebrates respond physiologically to bodily injury. The particular tissues injured also react and show various physiological changes at the local level. These adjustments are numerous, especially in the event of severe injury. One pertinent example is the response of the cortices of the adrenal glands in mammals that release corticosteroids in response to stress (Selye, 1950). This occurs under the stimulus of a rapid, substantial increase in the titer or circulatory adrenocorticotrophic hormone (ACTH), released from the anterior pituitary gland. High titers last possibly for only a brief period following injury in the newt (Liversage and Banks, 1980; see also Liversage and Korneluk, 1978; Liversage and Price, 1973). The various physiological responses to severe physical trauma ultimately affect the animal's metabolism as well as its electrolyte, water, and calcium–phosphate balances, which are altered in an attempt to cope with injury. Changes in the animal's physiology are greatly influenced by the hormones found to be essential for appendage regeneration in the adult newt, namely, insulin, growth hormone (GH), thyroxine (T_4), hydrocortisone (cortisol) (Vethamany-Globus and Liversage, 1973; Bromley and Thornton, 1974), and prolactin (PRL) (Connelly et al., 1968; Liversage et al., 1984).

Defect Experiments in Regeneration

On the basis of the findings of Walter (1910), Schotté (1926a), and Richardson (1940, 1945), hypophysectomy and thyroidectomy were initially the two most widely employed approaches in studying the role of endocrine secretions in adult newt appendage regeneration.

Defect experiments in earlier studies were appropriate for the following reasons: The pituitary controls various target endocrine glands (e.g., thyroid, adrenal cortices); defect experimental techniques were the basic approaches available, developed principally by the laboratories of the above-mentioned investigators; and replacement therapy of extracts from endocrine glands was not in common use, the purity of such hormones being questionable at best. Some incorrect interpretations have been made; nevertheless, a background of information has been realized, and new and better techniques and approaches designed. Not until the work of Stocum (1968a), using tissue culture methods, and the studies of Vethamany-Globus and Liversage (1973a) and Globus and Liversage (1975a,b), employing perfected organ culture techniques, were other than defect and transplantation experimental procedures available.

It is partly because of the limitation in available techniques that the approach to the study of hormones in regeneration became biased in favor of the search for the key, single regeneration hormone. The interaction of all the endocrine organs (the balance of the total Milieu) is affected by the removal of one endocrine organ, but this has not often been considered. Defect experimentation itself results in the tendency for investigators to search for the key hormone as opposed to considering the overall effect of the Hormonal Milieu.

Hypophysectomy

Concomitant hypophysectomy and forelimb amputation impedes regeneration but does not completely interfere with the early stages of epidermal wound healing, wound repair, and early dedifferentiation. However, if the hypophysioprivic condition continues (i.e., no replacement therapy), stumping ensues (Hall and Schotté, 1951). If a blastema is established before hypophysectomy, it continues to develop although growth and differentiation of the regenerate is affected, in some cases seriously, depending upon when hypophysectomy is performed after amputation (Schotté and Hall, 1952). In experiments by Schotté and Hilfer (1957), forelimbs were amputated through the regenerates concomitant with hypophysectomy. Amputated cone blastemas formed a small hand. When a palette or early digit stage regenerate (35–, 40– or 45–day regenerate) was amputated, it did not recuperate as readily and did not always continue to develop. Older regenerates (amputated after 45 or 60 days of regeneration) neither regenerated anew nor regressed but, rather, underwent stumping indicative of complete dependence upon hormones (i.e., the Hormonal Milieu), responding as if they were freshly amputated limbs. In this regard, the studies by Vethamany-Globus and Liversage (1973b) were concerned with the relationship between the anterior pituitary gland and the pancreas in adult newt tail regeneration. They found that under conditions of hypophysectomy, regeneration was abnormal and extremely retarded.

Thyroidectomy

Richardson (1940) obtained abnormal hindlimb regeneration in adult newts that were thyroidectomized 2 days prior to amputation and fixed after 9–15 weeks. Kambara (1953) injected thiourea, which interferes with the action of thyroid-stimulating hormone (TSH), into adult *Triturus pyrrhogaster* and found it inhibited tail regeneration. This effect could be counteracted by injecting thyroxine. However, when only thyroxine was added to the aquarium water containing hypophysectomized adult newts, survival and limb regeneration were poorly supported (Connelly et al., 1968). Indeed, when adult newt tail blastemas were cultured and treated with physiological (see Liversage and Korneluk, 1978) levels of thyroxine, limited growth and differentiation ensued (Vethamany-Globus and Liversage, 1973a).

If hyperthyroidism is induced in adult newts, the early regenerative stages are impeded (Richardson, 1945; Hay, 1956). Cartilage differentiation of the regeneration blastema in vitro is also limited in the presence of excess thyroxine (Vethamany-Globus and Liversage, 1973a, Fig. 11). These results emphasize the need for a balanced, physiological level of hormones. Thyroxine cannot be viewed as having an independent influence. The profile of the Hormonal Milieu as a whole is altered by either a hypo- or hyperthyroid state, and thus the interactions of the various hormonal components are out of balance. When Schmidt (1958a,b) studied the progress of forelimb regeneration in "short-term and long-term thyroidless adult newts" he found slight acceleration of regeneration several weeks after thyroidectomy. Regeneration may have been enhanced owing to reduced thyroid hormone levels. This implies that the thyroids normally tend to exert some control over the rate of regeneration and in their absence the earlier phases of regeneration are accelerated.

Other Hormone Systems

The adrenal bodies appear as small orange patches on the ventral surface of the kidneys of *Salamandra salamandra L.* and may also be found anterior to the kidneys as far as the subclavian artery, associated with the sympathethic ganglia (Francis, 1934). In an unsuccessful attempt, O. E. Schotté (1952, personal communication) once tried to remove all cortical tissue surgically in an adult urodele. Apparently, the islands are so numerous and diverse that to ablate the cortical tissue completely is very difficult. In addition, bleeding was extensive, not only clouding the surgical areas but also threatening the animal's survival. Bromley (1977) had some success ablating inter-renal tissue in the adult newt using electrocautery.

Schotté and Christiansen (1957) performed a series of experiments in another attempt to inhibit and thereby analyze the role of adrenocortical tissue in adult newt limb regeneration. Using the drug Amphenone B, which suppresses the synthesis of adrenal cortical hormones as well as thyroid hormones in mammals, the authors found that they could prevent forelimb regeneration in the treated adult newts. Their newts survived treatment and when the drug was withdrawn, the animals regenerated their forelimbs. Regeneration was restored in hypophysectomized newts when cortisone was injected together with Amphenone B, but not when ACTH was injected simultaneously with the drug.

Prior to 1973, no study had been made of the role of insulin in regeneration. At that time Vethamany-Globus and Liversage (1973a–c) investigated the effects of insulin on adult newt regeneration. In one series, they found that experimental diabetes caused by pancreatectomy resulted in a marked interference with limb and tail regeneration. In the cases in which a cauterized pancreas regenerated, there was a subsequent recovery of regeneration in the amputated appendages. Alloxanization also inhibited normal limb and tail regeneration. Such results suggest a role of insulin in the hormonal control of regeneration (Vethamany-Globus and Liversage, 1973c; Vethamany-Globus et al., 1978).

General Metabolism and Survival

As emphasized by Wallace (1981), the surgical removal of an endocrine gland affects the general metabolism (survival) of the adult newt, thereby possibly impeding regeneration. Animals do not survive beyond 27–30 days after hypophysectomy (Liversage and Liivamagi, 1971) and often die sooner. They exhibit many symptoms of hormone insufficiency, such as lack of muscle tone, and skin pigmentation changes. The epidermis does not molt, owing mainly to the depletion of the thyroid and adrenocortical hormones as well as prolactin (Kaltenbach and Clark, 1965; Dent, 1966; Schmidt, 1968; Hessler and Landesman, 1981a); consequently a cornified-like epidermal condition develops. The epidermis becomes abnormally crusty and black, mainly on the newt's dorsal side, due to the buildup of unmolted epidermis.

In experiments by Tassava (1983), the survival time and extent of forelimb regeneration in "well-fed" (overfed) adult newts was compared with that of fasted ones, following hypophysectomy. Intact newts were not fed for 3 weeks prior to hypophysectomy; they did not survive beyond 4 weeks after hypophysectomy, and regeneration was absent or abortive. However, newts well fed for 3 weeks before hypophysectomy generally survived up to 6 weeks posthypophysectomy (30% by 6 weeks) without food, and "many" regenerated to the digit stage. Tassava concluded that pituitary hormones are not an absolute requirement for forelimb regeneration in adult newts.

These results (see also Tassava, 1969) are interesting from the standpoint of the experimental design. The positive results obtained (regeneration) under otherwise negative experimental conditions (hypophysectomy) must, however, be related to overfeeding, which is likely a substitution for the natural condition, and therefore responsible for partial restoration of the Hormonal Milieu. Therefore we find the author's interpretations questionable. The following should be given consideration: (1) In "well-fed" newts, the enriched metabolic levels (e.g., blood glucose, glycogen stores, and purine nucleotides) and at least residual circulatory and somatic cell hormone levels might allow the Hormonal Milieu to remain sufficiently intact to partly support survival and regenerative activity following hypophysectomy and limb amputation (see Fig. 4.1). This would be comparable to the less than normal limb regeneration found in hypophysectomized newts given replacement therapy of embryonic chick pituitary extracts, as opposed to the nearly perfect regeneration of newts given more complete (quantitatively and/or qualitatively) therapy of adult chicken pituitary extracts (Liversage and Fisher, 1974). (2) Plasma corticosteroid levels of normally fed newts are not abolished (50% of control level) 10 days after concomitant hypophysectomy and limb amputation (Liversage and Price, 1973). (3) Innervated blastemas cultured in enriched medium (i.e., "well fed" and well maintained) display neither cell proliferation nor differentiation unless treated with hormones (GH or PRL with T_4, hydrocortisone and

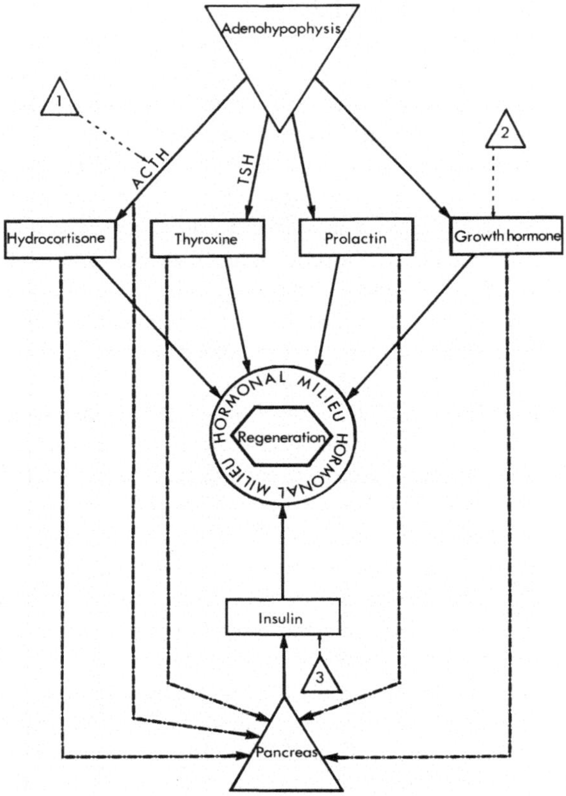

FIG. 4.1. Diagram of hormone interactions in appendage regeneration in the adult newt, modified from Vethamany-Globus and Liversage (1973c, Fig. 17) (△, source of hormone; ▭, hormone; ○, Hormonal Milieu; ⬭, process of regeneration; —, established relationships; - - -, possible additional pathways). Possible extraendocrine gland sources are lymphoid source of ACTH-like substance (△1; see Smith et al., 1982), nonpituitary source of GH-like substance (△2 ; see Rosenzweig et al., 1980), and nonpancreatic source of insulin (△3; see Rosenzweig et al., 1980).

insulin) (Vethamany-Globus and Liversage, 1973a; Liversage et al., 1984). (4) Several hormones and factors (e.g., ACTH, insulin, relaxin, somatostatin) have been isolated from nonendocrine cells in a variety of animals (Roth et al., 1982; Le Roith et al., 1983; Schwabe et al., 1983; Berelowitz et al., 1982). Although Le Roith et al. (1983) differentiate among endocrine, neurocrine (neurotransmitter), and paracrine (local, non-neural, or nonglandular action on neighboring cells) systems, they note that the molecules released are often identical among these systems. Such tissue might serve as a source of hormones (Fig. 4.1) assisting in the support of appendage regeneration in "well-fed" hypophysectomized adult newts.

Larval Amphibians

An exception to the findings of the hormone studies in adult newt can be found in larval urodeles. When kept in a hypophysioprivic condition, larvae readily survive and regenerate normal forelimbs (Schotté, 1926a, 1961; Liversage, 1967; see also Tassava et al., 1968). This is also true when *Ambystoma opacum* larvae (28–56 mm) are permitted to recuperate for 47 or 50 days between hypophysectomy and forelimb amputation (Liversage, 1967). This latter approach was considered desirable to test whether forelimbs of hypophysectomized larvae would retain their regenerative capacities following an extensive recuperation period. The results of the above investigations demonstrate that the presence of the anterior pituitary gland is neither requisite for normal limb regeneration in larval urodeles nor for the survival of the animal; such regeneration is hormone independent. Liversage (1967) also observed that growth and development of his experimental larvae continued normally, compared to control animals, up to the beginning of metamorphosis. All hypophysectomized larvae died, however, as metamorphosis began, indicating that various pituitary–endocrine interrelationships are essential for metamorphosis and continued survival. In addition, when larval *A. opacum* forelimb allografts (deplants) into the dorsal tail fin are innervated by an allografted nerve source and amputated and allowed to develop in vivo (Liversage, 1962) or in vitro (Liversage and Globus, 1977), the resulting regenerates give evidence of normal growth and differentiation in absence of hormones. Normal regeneration (both anatomical and physiological) also occurs when the spinal cord of larval *Ambystoma* is incised and allowed to undergo tissue regeneration in vivo (Butler and Ward, 1965). More recently, Horbay and Liversage (1977) transected the spinal cord in *Ambystoma* larvae and observed spinal cord regeneration in vitro in absence of hormones. Their findings support those of Butler and Ward (1965) in vivo.

Replacement Therapy and Regeneration

Thyroid Hormones

One of the first investigations in adult newts utilizing hormone replacement therapy was undertaken by Richardson (1945) using Antuitrin G and thyroxine. When adult hypophysectomized newts were injected (Richardson, 1945) with Antuitrin G (extract of GH, TSH, gonadotrophins) beginning 10 days postamputation, they regenerated better than newts that had been both hypophysectomized and thyroidectomized and subsequently given Antuitrin G. In the latter group, when both thyroxine and Antuitrin G were injected, more normal regeneration followed. Typical regeneration also occurred in thyroidectomized newts that received daily injections of thyroxine, beginning 10 days after amputation (see also Richardson, 1940). When Schmidt (1958c) implanted thyroxine pellets into thyroidectomized adult newts in the preblastemic stages, normal regeneration ensued. Pellet implantation during postblastemic stages invariably gave rise to stunted, abnormal regenerates. This work strongly suggests a stage dependence of growth and differentiation on thyroid hormones and is supportive of thyroid autografting results of Liversage and Brandes (1982). Moreover, pellet implantations were made near the amputation surface of a forelimb. The amputated contralateral forelimb exhibited more normal regeneration than the opposite forelimb containing the pellet. This suggests that the amputation surface closest to the pellet had an excess of T_4, whereas the contralateral regenerate received a more normal physiological level of the hormone(s).

Adrenal Hormones

On the basis of studies by Selye (1950) concerned with the "stress reaction" in mammals, Schotté (1953) carried out experiments and concluded that the ACTH fraction and various adrenocortical extracts he administered to hypophysectomized adult newts prevented premature invasion of the amputated forelimb wound surface by the dermis, "pointing to 'wound healing' as the most important phase in amphibian regeneration" and "the role of the (anterior) pituitary . . . as a regulator of special wound healing processes."

Schotté and Chamberlain (1955) also examined the role of the pituitary–adrenal axis. When daily injections of ACTH were given to hypophysectomized adult newts, near-normal forelimb regeneration ensued bilaterally in more than two-thirds of their cases, and the survival rate greatly improved. When these forelimbs were unilaterally reamputated through more proximal adult tissues and no ACTH replacement therapy given, regeneration was inhibited. The authors concluded that the (anterior) pituitary–adrenal synergism was reestablished when ACTH replacement therapy was given. In addition to this reestablishment, Liversage and Liivamagi (1971) and the present authors are convinced that the ACTH fractions employed by Schotté and Chamberlain (1955) were contaminated by other anterior pituitary hormones (see also Wilkerson, 1963, p. 223; Tassava, 1969). Thus, inadvertently, Schotté and Chamberlain provided their hypophysioprivic newts with at least a minimally adequate Hormonal Milieu, sufficient to sustain survival and also to promote and support regeneration. More recent findings by Hessler (1974) and Hessler and Landesman (1981b) show that regeneration will not ensue following hypophysectomy and treatment with relatively uncontaminated ACTH (see also Tassava, 1969).

In related experiments, Schotté and Bierman (1956) gave hypophysectomized adult newts exogenous preparations of various adrenocorticosteroids (eschatin, cortisone, and desoxycorticosterone acetate) in an attempt to stimulate forelimb regeneration. Fewer than half of their cases showed signs of regeneration, even fewer formed blastemas, and all animals died by 28 days postamputation. Unlike the fractions of ACTH used by Schotté and Chamberlain, the adrenocorticosteroids used in the latter experiments were probably uncontaminated by other hormones or only contaminated by other adrenal cortical hormones.

Insulin

It is now clear that injection of adrenocorticosteroids alone will not support regeneration in hypophysioprivic newts. It is quite possible that the "excellent regeneration in histologically proven absence of the pituitary" (Schotté and Bierman, 1956) was the result of cortisone stimulating or interacting with other hormone(s) (e.g., insulin; see Fig. 4.1). The relationship between hypophyseal and pancreatic activity has been shown by Vethamany-Globus and Liversage (1973b). Hypophysectomy adversely affected the normal histology of both the endocrine and exocrine parts of the pancreas, as evidenced by its atrophy. This suggested to the authors a possible involvement of the pancreas (beta cells producing insulin) in tail and limb regeneration in adult newts.

Growth Hormone and Prolactin

Niwelinski (1958) and Schauble and Nentwig (1974) showed that prolactin and growth hormone stimulate growth and differentiation of limb regenerates in normal adult newts. When exogenous prolactin is injected, maximum new growth occurs at 25°C, and maximum differentiation at 20°C (Schauble and Nentwig, 1974). Wilkerson (1963) induced forelimb regeneration in adult hypophysectomized newts with exogenous somatotrophin (80% somatotrophic hormone and 20% TSH and also bovine GH; see Fig. 4.2). He concluded that the absence of GH impeded regeneration (see Fig. 4.3). In order to obtain a full growth response in the vertebrates studied, the presence of the thyroid hormones and GH are necessary (Turner, 1966). Growth hormone alone has little effect on such target glands as the adrenal cortex, thyroids, and gonads, but it markedly enhances the effectiveness of the pituitary trophic hormones specific for those organs when administered with them, particularly in hypophysectomized animals (Li, 1956; Lostroh and Woodward, 1958).

Connelly et al. (1968), in follow-up experiments to those of Wilkerson (1963), showed that hypophysectomized newts exhibited 98% survival and

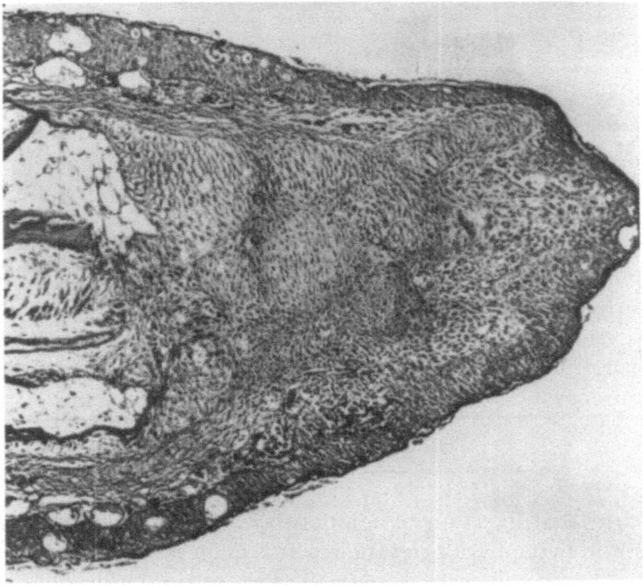

FIG. 4.2. Forelimb of a hypophysectomized adult newt that received 0.3 mg of bovine GH on alternate days for 15 days beginning at the time of limb amputation. The limb was fixed 29 days postamputation. A short, stunted, abnormal cartilaginous skeleton has formed and the epidermis is abnormally thickened, indicative of the hypothyroid condition (×60). (From Wilkerson, 1963, reprinted by permission of Alan R. Liss Inc.)

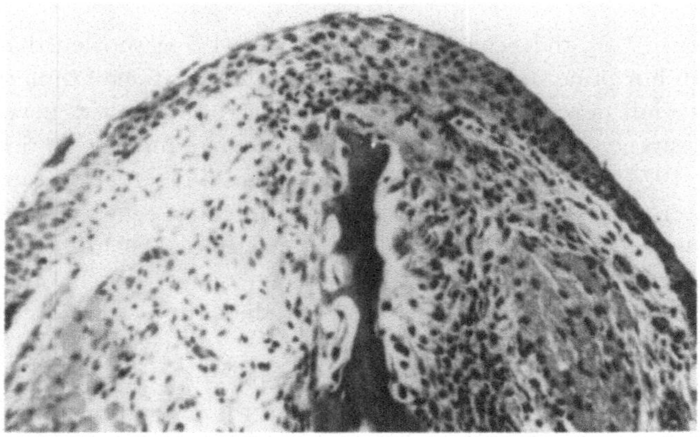

FIG. 4.3. Forelimb of a hypophysectomized adult newt 21 days after amputation. Regeneration was inhibited. The limb stump is covered by an epithelial cap and a thin fibrocellular tissue layer. Little dedifferentiation can be seen (×250). (From Liversage and Scadding, 1969, reprinted by permission of Alan R. Liss Inc.)

limb regeneration at 23 days posthypophysectomy (18 days postamputation) when injected intraperitoneally with PRL (National Institutes of Health, N.I.H. U.S.A), and kept in aquaria containing water with T_4 added. Prolactin alone increased survival and limb regeneration, but was less effective than the combination of PRL with T_4. Berman et al. (1964) have shown that mammalian PRL is a growth-promoting hormone in frog tadpoles. They point out that the bovine GH used by Wilkerson (1963) was contaminated with sufficient PRL (0.05 units/mg) to be an effective growth promotor had it been injected into *Rana* tadpoles in the dosage that Wilkerson used in his newts. As indicated above, PRL and PRL with T_4 support forelimb regeneration in hypophysioprivic adult newts. However, even the purified (N.I.H.) PRL used by Connelly et al. (1968) contained trace contaminants of growth, thyroid-stimulating, luteinizing, and follicle-stimulating hormones from the anterior pituitary gland. In short, in the preceding experimental approaches the following considerations should be kept in mind: (1) Both hormones injected (GH and PRL) were contaminated with other pituitary hormones; (2) all target endocrine glands were still present in the hypophysectomized animals; and (3) the presence of even trace contaminants could cause interactions between/among the injected compounds and the various endocrine glands, as well as among the endogenous glandular secretions, thus producing at least a minimal endocrine background. This series of events would begin the replacement of the normal Hormonal Milieu. On the basis of these considerations, we are not at all surprised that the survival of Wilkerson's and Connelly's animals was enhanced and that near-normal regeneration ensued with replacement therapy. Survival and regeneration are definitely not supported by any single "regeneration hormone" but, rather, a multiplicity of metabolic hormones of the Hormonal Milieu (see Fig. 4.1).

Hormone replacement therapy with GH or PRL, both alone or each in combination with T_4, has been shown to support survival and limb regeneration in hypophysectomized adult newts (Connelly et al., 1968; Tassava, 1969; Bromley and Thornton, 1974). Hessler and Landesman (1981a) injected hypophysectomized newts with PRL and TSH intraperitoneally in concentrations previously listed by Wilkerson (1963) as "contaminants" of his GH preparation. They found that regeneration occurred, suggesting that the animal's thyroids synthesized enough T_4 to aid in the support of regeneration. These results, in turn, indicate to us that a single hormone acts within the Hormonal Milieu in support of regeneration, despite the fact that the authors imply that one isolated hormone is responsible for regeneration. In subsequent experiments, Hessler and Landesman (1981b) injected hypophysioprivic adult newts on an alternate-day basis 8 days after forelimb amputation (13 days following hypophysectomy) with GH (N.I.H.) or PRL (N.I.H.), the latter in combination with T_4, and observed more or less normal regeneration. In the latter instance the authors described the "apical connective tissue of the limb stump undergoing a progressive erosion . . . [that] . . . became discontinuous."

They also referred to this "erosion" in regard to the findings of Wilkerson (1963) and Liversage and Liivamagi (1971, Fig. 7), who, Hessler and Landesman felt, "minimized the significance of the effect of hormone on the connective tissue pad in the non-regenerating limbs of their hypophysectomized newts prior to the administration of somatotropin or autoplastic pituitary implants, procedures which subsequently restored regenerative capacity to these limbs."

Hessler and Landesman (1981b) are of the opinion that these treatments must have resulted in the "erosion of the apical connective tissue present." Liversage and Liivamagi (1971) present no evidence supporting the hypothesis that erosion took place after resumption of limb regeneration. Following hypophysectomy in one series, they placed the pituitary into organ culture for 15 days. Two days after hypophysectomy the left forelimb was amputated. In absence of the pituitary, regeneration was inhibited; an incomplete fibrocellular pad and dermis invasion (note dermal skin glands) were evident in the regeneration area (see Fig. 7 in Liversage and Liivamagi, 1971). Following autoplastic implantation of the pituitary, a centrally located accumulation of blastema cells formed and could be seen flanked by the dermis and the incompletely formed cicatrix; regeneration then ensued (see center of regeneration area, Fig. 7, in Liversage and Liivamagi, 1971). Apparently, this blastema cell accumulation pushed the developing cicatrix and dermis aside (laterally) (see also Liversage and Scadding, 1969, Fig. 6). We do not subscribe to the idea that the cicatrix was "eroded."

The results of Landesman and Hessler (1981) reveal that one, two, or three injections of GH alone into hypophysectomized newts on successive alternate days resulted in digit stage regeneration by 30 days postamputation. Animals receiving PRL in combination with T_4 in a similar injection series merely underwent wound healing. From the results of their one to three injections of GH only, Landesman and Hessler (1981) conclude that the action of the hormone is essential during the early critical periods of regeneration and that it continues to have effects on the maintenance of the later regenerative processes. Apparently, only one to three injections of the PRL-T_4 combination does not support regeneration; PRL-T_4 injections must be given continuously (Hessler and Landesman, 1981b).

In mammals, GH stimulates growth by increasing serum amino acid and glucose titers, which leads us to speculate on an elevated glucose–insulin triggering mechanism. It also affects the rate of genetic transcription and translation, increasing protein synthesis (Martin, 1976). In addition, GH is involved in cyclic adenosine monophosphate synthesis, a second messenger in hormone-mediated events (Rothstein et al., 1980).

The findings from the laboratories of Landesman and Hessler support our view of the necessity for the Hormonal Milieu or combination of hormones in adult urodele appendage regeneration. When GH alone is given to hypophysioprivic newts (Bromley and Thornton, 1974), we believe it is the synergistic hormonal interaction, involving a number of endocrine secretions, that brings about a more complete Hormonal Milieu (see also Vethamany-Globus and Liversage, 1973c). An isolated hormone does not operate alone during regeneration (see Fig. 4.1).

Interspecies Hormonal Replacement Therapy

Experiments were designed to test the extent to which the forelimb of a hypophysectomized adult newt could regenerate given adult frog anterior pituitary extract (Liversage and Scadding, 1969). In studying the inter-relationship between the anterior pituitary and forelimb regeneration, we gave daily intraperitoneal injections of adenohypophyseal extracts from adult *Rana pipiens* (which do not regenerate limbs) to determine if this might promote normal limb regeneration and long-term survival in hypophysioprivic newts. Of 28 newts in the "biotest," 26 regenerated normal, advanced forelimbs. In a separate series, forelimb amputations were performed 5 days posthypophysectomy, but daily injections of extracts did not commence until 10 days postamputation (15 days posthypophysectomy). This was to determine whether regeneration could be re-established in a stumping forelimb of an animal that was in a state of endocrine insufficiency (see Liversage and Price, 1973). Liversage and Scadding (1969, Fig. 6) found retarded and abnormal regeneration in 10 of the 15 cases; the remaining 5 did not regenerate and exhibited dermal wound healing and ultimately stumping. The authors concluded from these experiments that normal forelimb regeneration and long-term survival of their hypophysioprivic newts was due to the presence of the continuous hormonal background provided by the replacement therapy.

Liversage and Fisher (1974) undertook additional biotest experiments involving intraperitoneally injected replacement therapy of adult or embryonic chicken anterior pituitary gland extract. Daily injections were not initiated until 7 days after limb amputation and hypophysectomy in the adult newts, in order to minimize circulatory titers of hormones. In 40 of 45 cases, limb regeneration ensued until fixation (25–30 days). In the embryonic chick pituitary series, however, regenerates were less advanced in their development. On the basis of the results of these experiments, the authors concluded that forelimb regeneration and survival of the hypophysectomized newts was due to the chicken pituitary extracts.

Our newt biotest system indicates that the frog and chicken pituitaries (hormones) used were not species specific. These results strongly support our conclusions that the situation within which the Hormonal Milieu acts is the determining factor. Although these hormones are not species specific with regard to their regeneration-promoting abilities, they are site or locale specific. The newt limb, at amputation, provides a structural and environmental area in which hormones can act to promote regeneration. This situation does not exist in amputated limbs of the frog and chicken donors.

Transplantation Experiments

Utilizing hypophysectomy and pituitary transplantation procedures, Schotté and Droin (1965) showed that following completion of the first metamorphosis in the developing newt, the anterior pituitary of the brown eft (first land form) does not influence limb regeneration, inasmuch as amputated forelimbs of hypophysectomized brown efts regenerate, as do limbs of unoperated brown efts. The presence of the pituitary, therefore, is not essential for limb regeneration even in the first postlarval stage, which is a condition similar to that of *Ambystoma* larvae (Liversage, 1967). According to Schotté and Droin (1965), the presence of the anterior pituitary is essential for limb regeneration only after second metamorphosis, or red eft stage (terrestrial), since the red eft, like the adult, does not regenerate following hypophysectomy. However, in the presence of an ectopically implanted adult newt pituitary gland, limb regeneration will ensue in a hypophysectomized red eft. The cell population of the developing pituitary gland increases, differentiates, and undergoes a gradual physiogenesis during the larval and brown eft stages (Atwell, 1921; Copeland, 1943; Schotté 1961; Schotté and Droin, 1965; Liversage, 1967).

From our earlier studies concerning the Hormonal Milieu in adult newt appendage regeneration, we hypothesized that a multiplicity of hormones is involved in regeneration and not an isolated, single "regeneration hormone." This thinking was borne out by our studies of stage dependence of hormones in regeneration, using autografting procedures.

Endocrine gland reimplantation experiments were undertaken in which the adenohypophysis of the adult newt (Liversage and Liivamagi, 1971) and later the thyroid glands (Liversage and Brandes, 1982) were surgically removed and placed into organ culture and subsequently ectopically autografted. In the adenohypophysis experiments, the glands were cultured for 21–24 days, whereupon they were autografted into the tail musculature or lower jaw. Left forelimb amputations were performed 2 or 3 days following pituitary extirpation. Right forelimbs were amputated 2 or 3 days after adenohypophysis reimplantation (between 23 and 30 days posthypophysectomy). Before pituitary reimplantation, left forelimb regeneration was thwarted. Organ culture, then autoplastic reimplantation of the adenohypophysis, permitted the authors to study the inhibition and later the concurrent restoration (left forelimb) and initiation (right forelimb) of regenerative activity, as well as the continuation of regeneration (see also Schotté and Tallon, 1960).

Liversage and Brandes (1982) further elucidated the role of the thyroid glands in adult newt forelimb regeneration. Following thyroidectomy, thyroids were placed into organ culture [according to procedures of Liversage and Liivamagi (1971)]. Forelimb amputation was performed 3 days postthyroidectomy. Glands were then autografted subcutaneously into the animal's lower jaw 9, 18, or 25 days postamputation. In this biotest of thyroid function, it was shown that when thyroids of an adult newt are autografted into an ectopic site

9 (preblastemic), 18 (blastemic), or 25 (postblastemic) days after forelimb amputation, recuperation, molting, and long-term survival of an animal will occur. The implanted glands also support restoration of forelimb regenerative activity. Only in the 9-day implant series, however, did normal growth, differentiation, and morphogenesis ensue.

Schotté and Lindberg (1954) tested their hypothesis "that an hyperadrenal state is necessary for regeneration" to occur. Their method involved xenoplastic transplantation of single or multiple frog (*Rana pipiens*, *R. palustris*) adrenals into hypophysectomized adult newts. Sham control animals that received frog muscle instead of adrenals did not show evidence of limb regeneration, while the experimental animals did regenerate, exhibiting "differentiation of skeletal formations." From our observations with concomitant hypophysectomy and forelimb amputation in adult newts, the first phases of regeneration (i.e., epidermal wound healing, wound repair, and initiation of dedifferentiation) occur. As endocrine insufficiency develops (Liversage and Price, 1973), proliferation of the dedifferentiated seed cells decreases and stops (Hoffman and Dent, 1977a,b) and characteristic stumping occurs. However, if 15 or 16 days after first amputation a single frog adrenal is transplanted into a hypophysectomized newt and the contralateral forelimb is amputated (Schotté and Lindberg, 1954), differentiation of skeletal elements ensues 23 days later. Two interpretations exist: (1) Possibly the foreign tissue implant brought about an immune response resulting in a stress to the animal and causing stimulation of the host's target endocrines (e.g., thyroids, pancreas beta cells) [Smith et al. (1982) provide evidence of "associations between the immune and neuroendocrine systems"] and/or (2) part of the Hormonal Milieu was restored by the frog adrenals and the remaining host endocrine tissues responded, owing to interactions among existing glands (see Fig. 4.1), as well as interaction between those glands and secretions from the frog adrenals. Vethamany-Globus and Liversage (1973a) have shown that blastemas in vitro treated with hydrocortisone alone do not grow and differentiate.

Schotté and Wilber (1958) hypophysectomized growing juvenile *Rana clamitans* frogs into which the adrenals from postmetamorphic *R. pipiens* were transplanted heterotopically beneath the jaw, concomitant with forelimb amputation. Regeneration was not pronounced. In a second series, 7 or 14 days after limb amputation, two to five extra *R. pipiens* adrenals were implanted into normal juvenile *R. clamitans*, thus providing the animals with excess adrenals. In the latter group 19 of 21 adult frogs showed evidence of regeneration that was probably not epimorphic but "tissue regeneration" (Carlson, 1970, 1978; Korneluk et al., 1982; Korneluk and Liversage, 1984) of the cartilaginous skeletal elements. Schotté and Wilber (1958) explained that "an artificially-induced hyperadrenal state is sufficient to determine recuperation of regenerative capacities in adult frogs." This "hyperadrenal state" was unproven, but, if present, it probably created a condition of artificial stress and a hormonal imbalance. Such a state might well have induced increased interactions among, and thereby increased secretion(s) from, the host frog's developing endocrine glands (e.g., thyroids, pancreatic beta cells, adrenal cortices, anterior pituitary). The hyperadrenal condition can be attributed not only to the presence of foreign tissues as a "stressor" (*R. pipiens* adrenals, and adhering kidney tissues in some cases), but also to the "low level of activity" (secretions) from the transplanted adrenals. This might be sufficient to induce a Hormonal Milieu in the host, capable of slowing the dermal invasion of the amputation site and also able to assist in supporting the proliferation of cells from the injured periosteum, apparently the main source of cells for this new development in anurans (see Carlson, 1970, 1978; Korneluk and Liversage, 1984). This work of Schotté and Wilber (1958) is probably the best example of extensively induced "tissue regeneration", in this instance of cartilage only, yielding a heteromorphic regenerate.

The stress of limb amputation in amphibians is responsible for ACTH release from the anterior pituitary, thereby stimulating increased synthesis and release of adrenocorticosteroids (glucocorticoids) (Schotté, 1961). In Schotté's opinion, the adrenocorticosteroids alone were essential for the initiation of the early stages of adult newt limb regeneration, and in their absence regeneration was inhibited and stumping occurred. However, when Schotté and Bierman (1956) gave adrenocorticosteroid replacement therapy to hypophysectomized adult newts following forelimb amputation, only a few animals formed small regenerates and large numbers died.

It was not until further investigations were undertaken that we could begin to conjecture that stress is involved at the time of amputation (injury), but that it is unnecessary for the more localized activity of limb regeneration (see Liversage, 1959); that is to say, the role of the stress of injury is probably necessary to aid in the animal's immediate survival and thereby its ability to cope with the stress of amputation (i.e., stress needed on the organismal level). However, the adrenocorticosteroids are not the only hormones essential for appendage regeneration at the local level. Insulin, GH, T_4, and hydrocortisone are essential (see Figs. 4.4 and 4.5), as shown by the studies of Vethamany-Globus and Liversage (1973a,c), as is PRL in combination with insulin, hydrocortisone, and T_4, as demonstrated by in vitro studies (Liversage et al., 1984) (see Figs. 4.6 and 4.7). The fact that these two combinations of hormones support regeneration emphasizes that these hormones operate in concert and that no one hormone should be singled out as the essential element.

Determinations of Hormone Levels

Liversage et al. (1971) measured the plasma corticosteroid levels in intact, control regenerate, sham hypophysectomized, and hypophysectomized adult killifish (*Fundulus heteroclitus*) 10 days after pectoral fin amputation. The data provided evidence for an anterior pituitary–adrenocortical axis in this fish, as

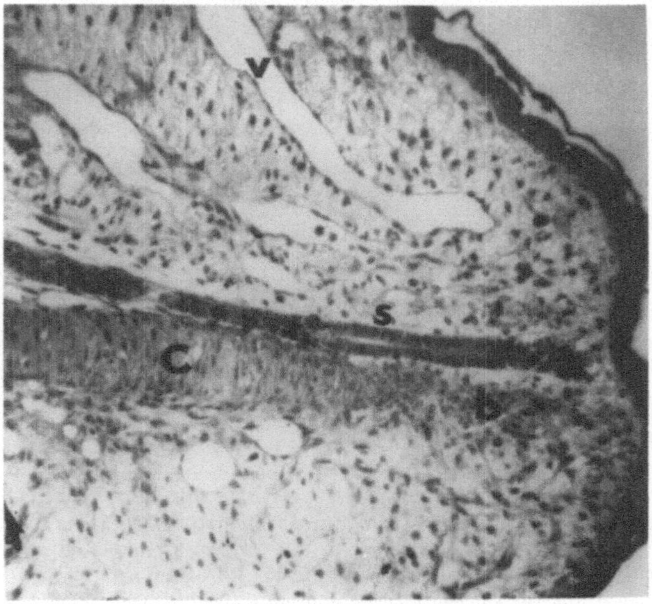

FIG. 4.4 Mid-sagittal section through a 15-day tail blastema explant from an adult newt cultured for 4 days with 0.2 μg of GH, 0.2 μg of hydrocortisone, and 5 μg of insulin per milliliter of culture medium. Advanced cartilage differentiation (c) has occurred, as well as extensive growth of the blastema and the regenerated spinal cord (s). Cell proliferation (b) is seen at the distal tip. Note the enlarged blood sinus (v). The arrow marks the level of amputation. (×100). (From Vethamany-Globus and Liversage, 1973a.)

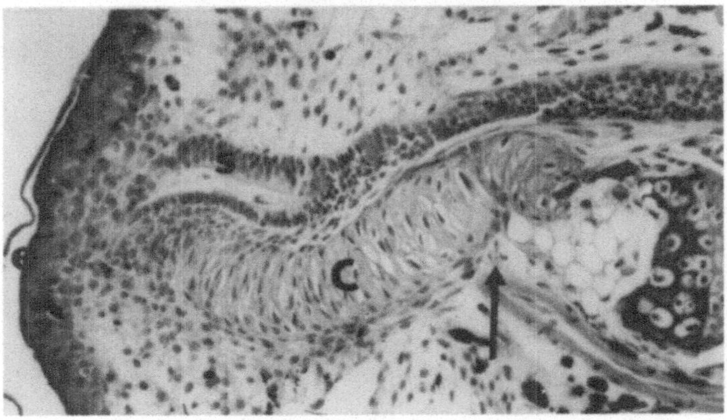

FIG. 4.5 Mid-sagittal section through a 15-day tail blastema explant of adult newt cultured 4 days with 5 μg/ml insulin, 0.2 μg/ml GH, and 10^{-8} mg/ml T_4. Note the extensive cartilage differentiation (c) ventral to the regenerated spinal cord (s). The epidermis (e) appears normal. The arrow marks the level of amputation (×120). (From Vethamany-Globus and Liversage, 1973a.)

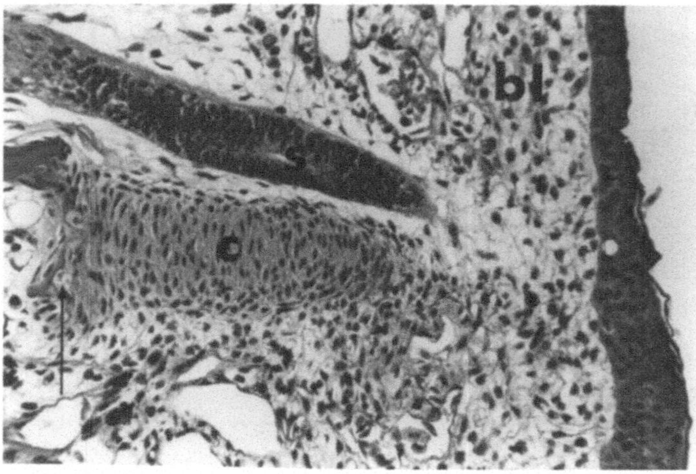

FIG. 4.6 Mid-sagittal section through a regenerate cultured with PRL-Sigma, hydrocortisone, insulin, and T_4. There is an appreciable mass of blastema cells (bl) present. Cartilage differentiation, as indicated by the cartilage bundle (c), extends distally, ventral to the regenerating spinal cord (s). Swirling patterns of cartilage cells are evident where the matrix has pushed and separated the cells during the process of cartilage development (×125) (From Liversage et al., 1984, reprinted from *Roux's Archives* by permission of Springer-Verlag.)

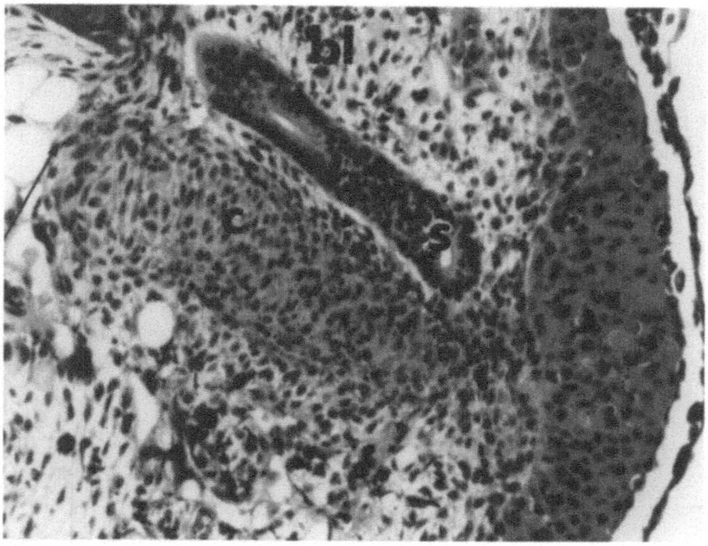

FIG. 4.7. Longitudinal section of a 13-day blastema cultured for an additional 96 hr in the presence of NIAMDD bPRL-6 (a relatively pure form of prolactin), hydrocortisone, insulin, and T_4. Similar to the regenerate in Figure 4.6 (treated identically, but with PRL-Sigma), as well as to control regenerates, the blastema cell population (bl) is near normal and cartilage differentiation (c) is progressing (×130). (From Liversage et al., 1984), reprinted from *Roux's Archives* by permission of Springer-Verlag.)

hypophysectomy was followed by a significant decrease in the plasma adrenocorticosteroid levels, although the fish survived and regenerated fins normally. Possible explanations are (1) the circulation and/or target endocrine, extra-, and nonendocrine cells may have contained residual sources of pituitary hormones that acted via target endocrines, as well as directly, to support survival and regeneration (see Roth et al., 1982; see also Fig. 4.1); (2) in absence of pituitary hormones, the pituitary's target glands may have continued to secrete residual titers of hormones; or (3) killifish fin regeneration may not be hormone dependent. On the basis of current evidence and findings in other vertebrates, the third possibility seems highly unlikely. More likely, a combination of the first two explanations existed over a period of time. Although normal fin regeneration ensued in hypophysectomized killifish (Liversage et al., 1971; Liversage, 1973) and the results parallel those of forelimb regeneration in hypophysectomized *A. opacum* larvae (Liversage, 1967; Liversage and Globus, 1977), the Hormonal Milieu appears to play a role in appendage regeneration in killifish, but not in *A. opacum* larvae.

Liversage and Price (1973) continued analytical investigations pertaining to the relative plasma adrenocorticosteroid levels in urodeles. They suspected adult newt forelimb regeneration of being glucocorticoid hormone dependent. These authors measured corticoid levels 10 days after adult newt forelimb amputation and hypophysectomy and found evidence for an anterior pituitary–adrenocortical axis. Their findings also showed that plasma adrenocorticosteroids are not completely abolished 10 days posthypophysectomy. When plasma samples were taken, the animals exhibited an abnormal physiological state. As the adrenocorticosteroid levels of nonhypophysectomized amputated, regenerating, and intact newts were within the same range, it appears that such regeneration proceeds in the presence of a relatively normal corticoid background, as postulated by Liversage (1959).

Liversage and Korneluk (1978) determined the concentrations of T_4 and triiodothyronine (T_3) in samples of fresh serum from regenerating adult newts, using a radioimmunoassay. Their findings showed that circulatory levels of T_4 and T_3 do not change significantly from levels in intact newts as regeneration progresses.

Liversage and Banks (1980) used a radioimmunoassay to determine the presence of insulin in the adult newt. No significant changes in the levels of insulin were found in newts during forelimb regeneration, compared with intact animals.

Bromley (1977) measured the preferential accumulation of ^{3}H-labeled exogenous cortisol and corticosterone in the "regenerating portions of adult newt limbs." He found glucocorticoid accumulation greatest 7 days postamputation. It was undetectable, however, following denervation in the preblastemic stage. When Bromley injected [^{3}H]corticosterone into hypophysectomized adult newts 4 and 15 days postamputation or into animals 1 day after their interrenal tissue had been ablated by electrocautery, the distal amputated segments showed accumulation of label. Bromley concluded that neither pituitary removal nor adrenalectomy affected the fate of injected labeled compounds. This accumulation effect was not observed in adult newts with 4-day regenerates, when other tritiated compounds (aldosterone, testosterone, cholesterol, estradiol, glucose, and leucine) were injected during normal regeneration.

The distribution of thyroid hormone during adult newt limb regeneration has been studied (Korneluk and Liversage, 1977) using radiolabeled thyroxine $[^{125}I]T_4$. After $[^{125}I]T_4$ injection, radioactivity was noted more in the regeneration area than in the proximal stump tissue and the contralateral forelimb. In a time course study of radioisotope accumulation (over 4 days), tissues at the blastemal stage showed an immediate increase in isotope, and the distal stump tissue a delayed accumulation. Other stages did not show as pronounced an uptake. This study lends support to the findings of Liversage and Brandes (1982) of stage-dependent regeneration in adult newt on thyroid hormone.

The influence of thyroid hormones during the proliferative stage of adult newt regeneration was investigated by McLaughlin and Liversage (1977). The findings revealed that 4 hr after injection of $[^{125}I]T_4$ and $[^3H]$thymidine, the animals showed preferential uptake of both labels in the 25-day tail and spinal cord blastemas. Thyroidectomized newts, 2 days postinjection, exhibited reduced $[^{125}I]T_4$ uptake in the blastemas compared with other tissue sites, although nonthyroidectomized newts showed the most label in the regenerates. The authors concluded that regenerating tail and spinal cord tissues accumulate thyroid hormone, especially during the proliferative phase of regeneration.

The fact that no change in the circulating levels of hormones occurs during regeneration and that regenerating tissue exhibits preferential uptake of the hormones demonstrates that any significant alterations with respect to the activity of hormones occurs in the local area of the regenerate. The aforementioned experiments indicate, to us, that the Hormonal Milieu affects regenerating tissue in a manner different from that in which it affects intact tissue.

Perspectives

The difficulty in deciphering the role of hormones in regeneration results primarily from (1) the tendency to consider hormones as individual entities, having a fairly autonomous role, once released into the blood stream, and (2) visualization of the hormonal target with little regard for its form and structure (see Fig. 4.8). Webster and Goodwin (1981) (see also Kauffman, et al., 1978; Newman and Frisch, 1979) have outlined a new approach to the study of morphogenesis. In their view, organisms are self-organizing structures, not reducible to atomistic elements. This structuralist viewpoint considers organisms to have control over their constituent parts through structural variations in their internal communication patterns. We consider that the zone of encounter between the surfaces of both the target tissue cells and the applied

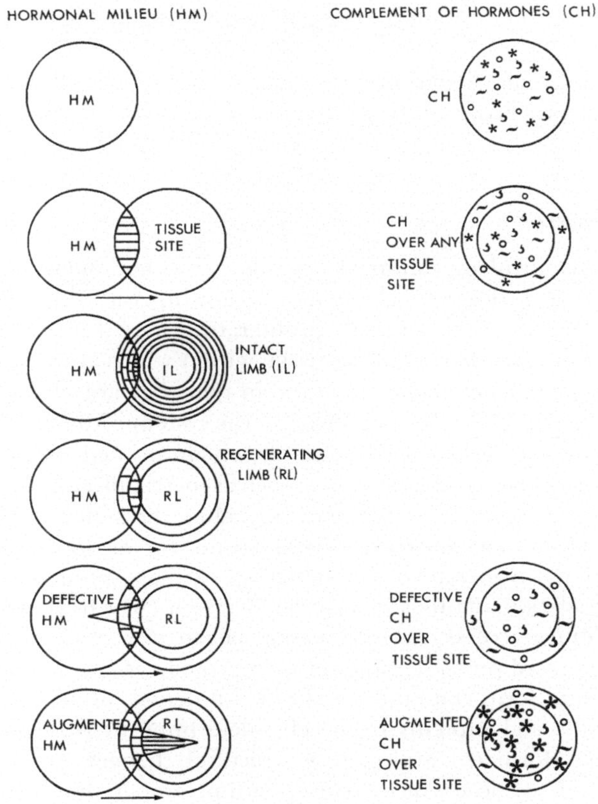

FIG. 4.8. Two models of approach to the study of hormones in regeneration. In the left column the Hormonal Milieu (HM) concept is depicted. The HM (circle) is viewed as a single entity. When the HM contacts a tissue site (another circle), time is taken (arrow) to penetrate into the tissue site. The congruence of HM with the tissue site leads to an altered condition (horizontal lines) for both the local HM and the local tissue site. The difference between the intact limb (IL) and regenerating limb (RL) is indicated in terms of accessibility, or number of boundaries (concentric circles) to HM activity. As the HM penetrates deeper into the tissue, it becomes more and more modified by the local site. The proximity of horizontal lines is directly proportional to the degree of modification. This HM–tissue interaction alters the activity of the HM through changes in both the HM and the tissue receptor sites. The HM, following a defect experiment, has its entire conformation and all of its inter-relationships altered. The defective HM (partial circle) presents its defect to the tissue site, and thus does not penetrate as effectively. An augmented HM is similarly presented as a circle, a whole entity, with an altered shape or profile. This augmented profile allows for deeper penetration by the entire new HM. In the right column, a simplified model of a more traditional view of hormone action in regeneration is presented. The Complement of Hormones (CH) (*ʖ ○ ~ in circle) consists of a collection of the component parts of the hormonal background. Here the overall shape or profile is not considered. The interactions of CH and the tissue site is not considered with regard to the structure or locale of the tissue; CH action is a simple superimposition of the CH onto the tissue site (a concentric circle). Variations in receptor sites are the only control of hormonal action. A Defective CH is one that is missing a component part, a loss which has little effect on the activity of the other constituents. An augmented CH has a surplus of one or more component parts, again having little effect on the overall profile of the CH.

hormonal background determines the type of activity that ensues, rather than do individual properties of either the tissue cells or the Hormonal Milieu alone.

Tissue Structure

Morphogenesis involves continual episodes of containment and isolation. Tissue sites have a constantly changing relationship with their external environment. Cells that are in a rapidly proliferating state, for example, create a cell accumulation and then a bulk of tissue within which are small volumes of intercellular space. These small volumes, in that they are physically isolated, are resistant to the relatively rapid fluxes in the systemic Milieu. Also, salts and proteins, which such cells extrude, tend to remain in their vicinity for longer periods of time. Other structural events, such as matrix production, also change the access of the external environment (including the systemic Milieu) to the tissue sites. Reddi and Kuettner (1981) found, for example, that protease inhibitor activity in mammalian noncalcified cartilage produced tissue resistance to degradation and thus to vascular invasion (which would provide the Hormonal Milieu access). As calcification began just prior to osteogenesis, there was decreased protease inhibitor activity and vascularization proceeded.

The structural form and other local environmental influences, such as the leakage of salts and proteins from cells, determine in part the access and profile of the Hormonal Milieu. Any structural changes that take place in tissue sites, such as those which occur upon limb amputation, must be considered as controlling factors during regeneration. To tissues that have been cut from their former tightly contained structure and matrices, subjected to degradative enzymes and osteoclasts, and placed in hitherto unlikely juxtapositions with one another, the Hormonal Milieu will take on a different profile and will have different effects.

Form and Derivation of the Hormonal Milieu

The Hormonal Milieu itself must be viewed as an entity (see Figs. 4.1 and 4.8), and not as a composite of several autonomous items. In effect, the coming together of the surfaces of two forms, the target cells and tissues and the Hormonal Milieu, is that which stimulates a particular response in the regenerate tissues. This response can be modulated at the level of the receptor sites in the target tissue cells. Equally effective, however, would be a change in the shape of the Hormonal Milieu profile.

The Milieu in the vascular system of an animal is not delivered in a uniform fashion to all tissue sites. In a clump of cells, the small intercellular spaces provide a contained, sheltered environment in which salts and proteins collect. This environment can have an effect on the pH, the osmolarity, and the activity of the hormones. Even more exaggerated local change would be provoked by in situ hormonal production. Recently, the highly defined roles and sites of production for hormones have been challenged (Roth et al., 1982). Peptides very similar to hormones and other messenger molecules of vertebrates have been discovered in extracts of unicellular eukaryotes and prokaryotes (Le Roith et al., 1983; Berelowitz et al., 1982; Schwabe et al., 1983). The presence of these hormones across the phylogenetic scale suggests that they perform a general function in intercellular communication and that it is the local microenvironment that dictates the specificity.

Sandor and Mehdi (1982) observed that the biological messenger system is highly conservative and that the endocrine system resulted from the "opportunistic" evolution of target organs. The presence of hormonelike substances in lower forms indicates that any individual hormone may play a variety of roles and is subject to its local conditions. Moreover, Rosenzweig et al. (1980) performed experiments in mammals which suggested that extrapancreatic tissue insulin might be synthesized by the nonpancreatic tissues themselves. They conclude that such locally produced insulin might be involved in activities other than carbohydrate metabolism.

Evidence for other locally produced hormones is also available. Human chorionic gonadotropin, human growth hormone, cholecystokinin, and vasoactive intestinal polypeptide (Yoshimoto et al., 1977; Rosenzweig et al., 1980; Krieger, 1978) have been detected in ectopic sites. Significant levels of TRH, TSH, PRL, and ACTH have been demonstrated in the brains of hypophysectomized rats (Krieger, 1978). Smith et al. (1982), studying hypophysectomized mice infected with Newcastle's disease virus, found an increase in corticosterone production. They suggest the existence of a lymphoid–adrenal axis, since the virus stimulates the release of a lymphoid-derived substance related to or possibly identical with ACTH (see Fig. 4.1 and Le Roith et al., 1982). It is entirely likely, therefore, that these and other hormones (or hormonelike substances) produced by cells outside the conventional endocrine glands exert influences on target tissues that are not traditionally associated with these hormones.

Interaction of the Hormonal Milieu and Injured Tissue

The regenerating adult newt limb can be described as a new structure, one that is considerably altered from the intact limb. During the early stages of regeneration, mesodermal cells lose their connections to the differentiated tight matrices and array themselves in a loose conglomeration of embryonic-like cells (Butler, 1933, 1935). The Hormonal Milieu exerts its influence during regeneration, not by altering its own composition and allowing one or another single hormone to dominate, but in that it obtains greater access to the injured and exposed tissues as well as to the cells and tissues of the regenerate. The altered profile is mediated by structural change.

Both qualitative and quantitative changes in hormone receptor sites can be postulated; however, we consider that these receptor site variations are the result of the breakdown of differentiated tissue and the accumulation of the loose array of blastemal cells. The increased access that the Hormonal Milieu has to these cells can lead to the increased local uptake of hormones. Hormonal titers, exceeding normal levels (e.g., replacement therapy experiments), might be stimulating low-affinity receptors in the regenerating tissue; any one hormone, in larger quantity, might, in addition to its own sites, compete for the receptors of other hormones. Similarly, following defect experimentation, other endogenous hormones may compete for the receptor sites of the missing hormone(s). Such conditions would probably lead to a cell acquiring an excess of one hormone and a deficiency in other hormones (see Fig. 4.8). This would yield unpredictable results.

The fact that injury itself or injury-induced local environmental change can be instrumental in the alteration of the Hormonal Milieu is illustrated in our experiments that measured the level of cyclic nucleotides in adult newt limb regenerates and denervates (Liversage et al., 1977; McLaughlin et al., 1982; 1983). Cyclic nucleotides are involved in regeneration, more specifically the Hormonal Milieu, in that they act as second messengers in hormone-mediated events for a number of systems including serum hormones (see Robison et al., 1971). It should be noted that cyclic nucleotide production would be especially susceptible to fluctuations in ion levels for example (McMahon, 1974), produced by local sources and by local structural conditions. In our experiments, it was found that amputation and/or denervation led to significant fluctuations in the levels of cyclic nucleotides, not only in the operated limb but also in a parallel fashion in the contralateral intact forelimb. This local change, which is not evident in the hindlimb, is possibly transmitted via nerve axons, as there is nerve damage during both amputation and denervation. The fluctuations we observed for the 72 hr following denervation were similar to the changes in DNA, RNA, and protein levels following denervation (Lebowitz and Singer, 1970; Singer and Caston, 1972; Geraudie and Singer, 1978); however, our sham denervations of nonamputated limbs led to fluctuations that paralleled, although in a somewhat delayed fashion, those of the denervated limbs. Some damage may occur to nerve axons during sham denervation; however, it is also true that the newly exposed injured tissue presents a novel structure. Also, sham denervation led to cyclic nucleotide level alterations in the contralateral limb, which again mimicked the results obtained after denervation.

In our view, the Hormonal Milieu is a structure or form in its own right. This form functions in a way that relates to its shape or profile and to the profile or receptivity of the target organ. The profile of the target organ, partly based on its structure, can affect the shape of the Hormonal Milieu locally. Areas allowing little access will lead to the depletion of certain hormonal elements. Thus the Milieu profile changes inside such a tissue site. Local production of ions, proteins, and possibly hormones, will alter the Milieu structure as well. Experiments involving removal of one hormone-producing gland or another cannot be regarded as isolated events. The removal of one hormone affects the profile of the total Hormonal Milieu, and thus affects the interaction of the other component parts. The new form of the Milieu will interact differently with regenerating tissues. The failure of a limb to regenerate normally can be seen as the result of a change in the total background Milieu, rather than as a depletion of one vital component (see Fig. 4.8).

This approach provides a new outlook concerning the action of hormones in regeneration. Rather than as a collection of factors, the Milieu can be experimented upon as an integral, albeit diffuse, organ in its own right. Experiments can be concerned with the stability of the Hormonal Milieu. Stability, in our terms, is defined as an object's ability to maintain its differentiation from surrounding objects. Defect experimentation, by altering the profile of the Milieu, can be said to have disrupted this stability. However, developmental or at times even accidental or experimental change should not adversely affect the long-term stability of an object, as long as a differential between one object and another is maintained. The Hormonal Milieu, seen as an object, can reach new stable conformations depending upon local conditions. There are, most likely, unstable states of transition between stable conformations (see Thom, 1975). Removal of one or more hormones from the Milieu may result in unstable conditions, leading to a lack of integrity of the Hormonal Milieu and to the eventual death of the organism. It is equally possible that the removal of one element could result in a new stability, although not necessarily the most optimal with regard to regeneration. The importance of the hormonal background as an active participant in regeneration will become more obvious as experimentation becomes concerned with the Hormonal Milieu as a discrete object rather than as a collective aggregate.

Interaction of the Hormonal Milieu and the Genome

A clear understanding of a cell's microenvironment, and secondarily of the general systemic environment of an organism, will allow greater possibility for experimental control of new development. Regeneration would be induced (e.g., via gene de-repression by the Hormonal Milieu) in an otherwise nonregenerating organism. There are indications that the genome has the same basic constitution in terms of coding sequences whether a cell is developing embryonically, is differentiated, or is involved in the regeneration of an adult newt appendage; only the expression of the genome varies during the different phases of development, not the genome itself (see Gurdon, 1974; Briggs and King, 1952, 1957).

The genetic manifestations exhibited by a cell are mainly a matter of expression, repression, or de-repression of the genes. This results in morphological alterations to the cytoplasm and, as a consequence, to changes in the immediate or microenvironment surrounding the cell as it changes shape and size and acquires a tight structural organization. During development these cytoplasmic constituents undergo profound changes leading to cellular specialization. In short, the environment of a cell, including the Hormonal Milieu, is a major factor determining whether regeneration will ensue.

Summary

During the normal course of regeneration of an adult newt appendage, the four major physiological factors that are considered essential for new development are amputation injury, epidermal contact, nerves, and hormones, all of which, in concert, play a role in epimorphic regenerative activity.

The "multiple hormonal background," from which we have derived the more unified concept of the Hormonal Milieu, is involved in growth and differentiation in adult newt appendage regeneration. The new structure and environment, established during the early stages of regeneration, are the framework upon which the Hormonal Milieu acts. In turn, the environment in which hormones operate, in part created by the hormonal action itself, determines the activity of the Hormonal Milieu. The term *Hormonal Milieu* distinguishes the hormone background as an integral entity within the organism; hormonal systems interact with one another and their surrounding local (micro-)environments to form a larger interrelated system, the Milieu. The Hormonal Milieu exerts its influence during regeneration, not by altering its own composition and allowing one or other single hormone to dominate, but by obtaining greater access to the injured and exposed tissues as well as to the cells and tissues of the regenerate. The influence exerted upon a cell by its microenvironment, which includes the Hormonal Milieu, is a major factor or key influencing and determining whether regeneration will ensue.

A model of a more traditional view of hormone action in adult newt epimorphic regeneration has been presented and compared with our current Hormonal Milieu model.

5 Neurotrophic Influences on Cellular Proliferation in Urodele Limb Regeneration: In Vivo Experiments

ROY A. TASSAVA AND CHERIE L. OLSEN

Importance of Cellular Proliferation for Epimorphic Regeneration

A period of cell division is a prerequisite for epimorphic regeneration to occur, be it deer antlers (Goss, 1972), tissue punched out of rabbit ears (Grimes and Goss, 1970), the extirpated newt lens (Reyer, 1954), or amputated urodele limbs (Thornton, 1968; Carlson, 1974a). Cellular proliferation provides the cell population, the blastema, from which the new tissue can be constructed by differentiation, morphogenesis, and pattern formation (Hay and Fischman, 1961; Bryant and Iten, 1976; Tank and Holder, 1981).

In general, regenerating tissues are composed of asynchronously cycling cells; that is, not every cell undergoes division at the same time. It is now recognized that blastema cells, like other proliferating cells, pass through what is called the cell cycle, a period of time required to go from one mitotic division to another, and that the important regulating events occur in the "interphase" period between mitoses (Mitchison, 1971; Tassava and McCullough, 1978).

There are four phases to the blastema cell cycle; the *M phase*, or mitosis/cell division period; a gap, the *G_1 phase*, between mitosis and the *S phase* (the DNA replication period); and a gap, the *G_2 phase*, between the completion of the S phase and another mitosis. The various times for these phases and the total cell cycle times (CCT), as measured by various investigators for regenerating urodele limbs, are given in Table 5.1. It is important to note that the M phase is very short, about 1 hr, compared to the S phase, which is over 30 hr. Thus more cells of a regenerating limb blastema can be found in the S phase than in the M phase at a given time. Cells in the M phase can be recognized as mitotic figures (prophase, metaphase, anaphase, and telophase). Cells in the S phase can be recognized by use of autoradiography, because they will incorporate [^{3}H]-thymidine ([^{3}H]T) into replicating DNA, that is, they become labeled. Because cells synthesize RNA and protein throughout the G_1, S, and G_2 phases, the only way to distinguish between G_1 and G_2 cells is by microspectrophotometry; G_2 cells have the 4C content of DNA, G_1 cells have the 2C quantity, but neither is replicating DNA.

TABLE 5.1. Duration of Cell Cycle Phases in Regenerating Urodele Limbs

		Cell cycle Parameters (hr)					
Blastema type	Series	G_1	S	G_2	M	CCT	Reference
18-day newt blastema	No chase	2.5[a]	41	4.0[a]	1	47.5	Grillo (1971)
21-day newt blastema	1000× chase[b]	4.6	32	5.1	1	42.7	Laux (1980)
8- to 9-day axolotl blastemas	100× chase	3.2	40	4.8	1	49	Wallace and Maden (1976)
8- to 9-day axolotl blastemas	1000× chase	9.5	38	4.5	1	53	Wallace and Maden (1976)
9-day axolotl blastemas	No chase	1.9	32	5.1	1	40	McCullough and Tassava (1976)
6-day axolotl blastemas	No chase	4.5	34	5.5	1	45	Whiteside (1978)

[a]Values recorded in the literature as $G_1 + \frac{1}{2}M$ and $G_2 + \frac{1}{2}M$; 0.5 hr was subtracted from these values to arrive at G_1 and G_2 values.
[b]Here 1000× chase refers to an injection of unlabeled (cold) thymidine (1000× greater by weight than the dose of [^{3}H]T) given 2 hr after the [^{3}H]thymidine to ensure a "pulse" labeling of only those cells in the S phase.

If it is assumed that every cell in a blastema of a regenerating limb is progressing through the cell cycle, then the following relations hold true:

$$\text{mitotic index (MI)} = \frac{\text{number of cells in M}}{\text{total number of cells}} \times 100 = \frac{\text{length of M}}{\text{length of cell cycle}} \times 100$$

and

$$\text{labeling index (LI)} = \frac{\text{number of cells labeled (pulse of } [^3\text{H]T)}}{\text{total number of cells}} \times 100 = \frac{\text{length of S}}{\text{length of cell cycle}} \times 100$$

It can be seen that the mitotic index will rarely surpass 3%, whereas the [^{3}H]T labeling index can theoretically approach 70–80% (see Table 5.1).

The cells of regenerating limb blastemas originate via dedifferentiation of mesodermal stump tissue cells in the vicinity of the amputation (reviewed by Hay, 1966b; Thornton, 1968). Critical experiments demonstrating the source of blastema cells were done by Hay and Fischman (1961) and Steen (1968), the results of which showed that the cells of the blastema come from muscle, connective tissue, and cartilage.

Before a limb of an adult newt is amputated, cells of the mesodermal tissues are not proliferating. The MI and LI are near zero and, in fact, do not begin to increase until day 3 (DNA replication begins by some cells) and day 4 (mitosis) postamputation, correlated with the beginning of dedifferentiation (Chalkley, 1954; Hay and Fischman, 1961). RNA and protein synthesis show increases as early as 2 days postamputation (Bodemer and Everett, 1959; Kelly and Tassava, 1973). Continued dedifferentiation, along with some early cell divisions, results in an increase in the number of proliferating cells in the distal limb stump. In adult newts, the MI and LI increase through 2 weeks postamputation and a blastema develops concurrently (Chalkley, 1954; Hay and Fischman, 1961; Mescher and Tassava, 1975). In small *Ambystoma* larvae, blastemas form as early as 5–6 days postamputation (Schotté and Butler, 1944; Stocum, 1979). As the blastema enlarges and protrudes distally beyond the amputation surface, the LI and MI increase to their maximum values and then decrease as cells leave the cycle from G_1 to differentiate (Singer and Craven, 1948; Chalkley, 1954; Hay and Fischman, 1961; Tassava and McCullough, 1978). Differentiated cells generally have a 2C DNA content and are said to be in G_0.

Various drugs that inhibit RNA, protein, and DNA synthesis, or mitosis, would be expected to inhibit regeneration, and various studies have shown this to be true (reviewed by Carlson, 1974a; Wallace, 1981).

Influence of Nerves and Other Factors on Cell Proliferation During Limb Regeneration

A central problem in limb regeneration is concerned with what factors initiate and maintain cell proliferation. Factors present in serum or released from injured, degenerating nerve may play roles (Mescher and Gospodarowicz, 1979; Mescher, 1982). Because the wound epidermis is so essential to limb regeneration, the possibility is likely that a chemical interaction occurs between wound epidermis and underlying stump cells (Singer and Salpeter, 1961; Thornton, 1968; Stocum and Dearlove, 1972; Mescher, 1976). Various hormones have been investigated with regard to possible influences on limb regeneration (reviewed by Thornton, 1968; see also Chap. 4), including prolactin, growth hormone, thyroxine, adrenal steroids, and insulin (Schotté, 1961; Bromley, 1977; Vethamany-Globus and Liversage, 1973a–c; Vethamany-Globus et al., 1977). No in vivo experiments have demonstrated a direct action of hormones on the cells or tissues of the regenerating limb (Tassava, 1969, 1983).

The most frequently investigated factor has been the putative neurotrophic factor (NTF), a factor that has been elusive to identify and purify (see Chap. 6

for a more detailed discussion), given the considerable amount of evidence in support of its existence. Since the early work of T. J. Todd, Schotté, Butler, and Singer (reviewed by Singer, 1952, 1974), it has been known that urodele forelimbs do not regenerate if denervated.

There are three basic in vivo experimental approaches designed to investigate the role of nerves in limb regeneration: (1) The forelimb is denervated at some time prior to amputation (predenervated limbs), (2) denervation is on the day of amputation or 1 day postamputation, or (3) denervation is delayed by some days or weeks after amputation. Several important observations have emerged from these experiments, the most important being that cellular proliferation is severely affected by denervation. When forelimbs of *Ambystoma* larvae are amputated and simultaneously denervated, a typical blastema does not form at the limb tip. Histolysis of stump mesodermal tissues does occur and the few dedifferentiated-like cells that appear at the limb tip have been referred to as a pseudoblastema (Butler and Schotté, 1941; Olsen and Tassava, 1984). Dedifferentiation also initially occurs in amputated, denervated limbs of adult newts (Mescher and Tassava, 1975), but a pseudoblastema is not seen. Several layers of flattened cells later form at the distal limb tip (Salley and Tassava, 1981).

Only a minimal amount of cell division has been observed in denervated limb stumps of *Ambystoma* larvae (Kelly and Tassava, 1973; Tassava et al., 1974; Maden, 1978, 1979a) or of adult newts (Mescher and Tassava, 1975). The observations that cells dedifferentiate, increase their rate of RNA synthesis (Kelly and Tassava, 1973), and exhibit a slight increase in [^{3}H]thymidine LI (Mescher and Tassava, 1975; Maden, 1978) suggest very strongly that a denervated amputated limb makes an attempt to regenerate but, without nerves at the distal limb tip, a blastema never develops. Since a blastema forms as a result of cellular proliferation, it can be concluded that one or more events in the series of events necessary for cell division does (do) not occur in a denervated limb (see Garling and Tassava, 1984).

Results of experiments that involve delayed denervation support the above conclusion, but here the situation is more complicated. Young blastemas progress through a nerve-dependent period during which regeneration comes to a halt if the limb is denervated. Blastemas of the early to mid bud stage, that is, to approximately days 7–8 in small *Ambystoma* larvae and to days 13–15 in adult newts (Tank et al., 1976; Stocum, 1979; Iten and Bryant, 1973), stop developing after denervation and resorb back to the level of amputation. Older blastemas, of late bud, palette, or digit stages, continue to develop after denervation, but the final volume of the regenerate is adversely affected (Schotté and Butler, 1944; Singer and Craven, 1948).

Singer and Craven (1948) examined mitoses per unit volume of blastema in adult newts after denervation at nerve-dependent and nerve-independent stages (see Fig. 3, p. 291, in Singer and Craven, 1948). Interestingly, denervation at nerve-dependent stages has a more adverse effect on mitotic activity than at later nerve-independent stages. These results will be discussed below in relation to the cell cycle.

The Nerve Threshold Theory

For regeneration to occur, it has been hypothesized that a critical or "threshold" number of nerve fibers must be present in a urodele forelimb (reviewed in Singer, 1952). Singer and his colleagues have determined this number for adult newt forelimbs to be just above one-third of the total number of nerve fibers normally innervating the forelimb. By cutting various combinations of the motor and/or sensory roots of the 3rd, 4th, and 5th spinal nerves, Singer (1952) showed that as long as the number of nerve fibers (or axonal area; see Singer et al., 1967) is above the threshold number, either motor nerves alone or sensory nerves alone can support regeneration. Later, Kamrin and Singer (1958) reported regeneration of denervated newt limbs after implantation of dorsal root ganglia into the distal limb stump. Again there was a correlation between the number of ganglia implanted and the percentage of limbs showing regeneration.

It is clear from these experiments that nerves exert their trophic effect in the local environment of or directly on the cells participating in regeneration. In fact, the possibility has been raised that nerve fibers actually have to contact dedifferentiated/blastema cells to stimulate their proliferation (Salpeter, 1965; Tassava and McCullough, 1978).

There are two alternate ways of visualizing the nerve threshold requirement at the cellular level. If it is assumed that a neurotrophic factor (perhaps a protein) is synthesized in the nerve cell body and transported down the axon, then it could, as first alternative, be elaborated into the limb stump/blastema from the tips of regenerating axons (neurites). If the NTF is readily diffusible, then every cell in the distal limb stump/blastema would have equal access to it. In this case, a threshold requirement of nerve fibers would be necessary to ensure a minimum concentration of NTF in the distal limb stump/blastema. Below this limiting concentration cells could not proliferate (see Fig. 5.1).

According to the second alternative (Fig. 5.1), the NTF is delivered by direct (or close) contact of neurites and the cells of the regenerate. This method of delivery would ensure a trophic effect of a labile chemical before its breakdown (Salpeter, 1965; Globus, 1978) or a trophic effect of a neurite membrane-bound factor. A trophic (mitogenic) effect of a rat neurite membrane fraction on Schwann cells has been demonstrated by Salzer et al. (1980b). With the second alternative, it is not as easy to account for a threshold requirement. As long as a few cells have neurite contacts, some cell division should occur and regeneration should progress, however slowly. Partial denervation does have two adverse effects on regeneration: The process is delayed and the regenerates are smaller than controls (Singer and Egloff, 1949; Tassava et al., 1982). Perhaps

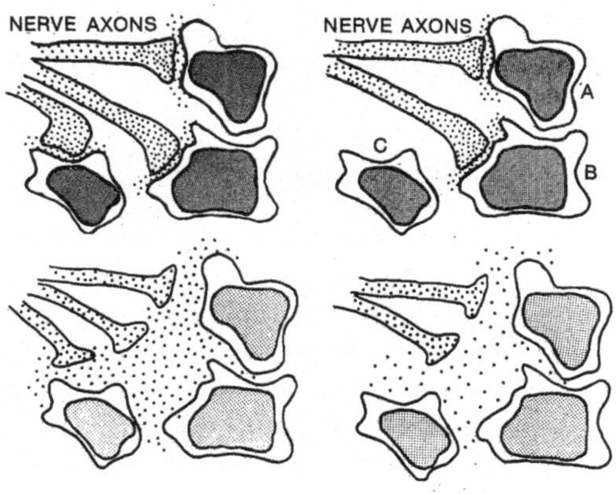

FIG. 5.1 Diagrammatic illustrations of two ways to visualize the nerve threshold concept in regeneration. In the two upper drawings, nerve fibers must make close contact with blastema cells to impart the neurotrophic stimulation. If a cell is not contacted by nerves (upper right, cell C), it does not proceed through the cycle. In the two lower drawings, nerve fibers release the neurotrophic factor into the blastema milieu such that each blastema cell has equal access to it. When the concentration falls below a critical level (lower right), cell cycling stops.

below a certain number of nerve contacts more cells are stopped (blocked) in the cell cycle than proceed. If the blocked cells die and are removed from the population, the few cells that divide provide an insufficient number of new cells to form a blastema. The minimum number of nerve fibers needed to obtain a net gain of cells would then be the threshold level.

The Cell Cycle and Neurotrophic Controls

The experiment reported by Kelly and Tassava (1973) (see above) stimulated interest in the examination of control factors and the cell cycle in limb regeneration. It was clear that a blastema does not form on a denervated limb stump, but Kelly and Tassava (1973) reasoned that many molecular/cellular events must occur in innervated limbs prior to blastema formation and asked which of these events, if any, occurred in denervated limbs. Kelly and Tassava (1973) compared MI and RNA synthesis in innervated and denervated limb stumps of *Ambystoma mexicanum* larvae. Tassava, his colleagues, and other investigators have continued these kinds of comparisons, the results of which are shown in Table 5.2.

It is obvious that the main difference is the absence of a sustained period of DNA replication and mitosis (cellular proliferation) in denervated limbs. The cells in denervated limb stumps appear to initiate some events in the direction of regeneration, as indicated by an early increase in DNA labeling index (LI) and mitotic index (MI), although neither is sustained. At least some of the RNA and protein synthesized in denervated limb stumps therefore appears to be directed toward cell cycling (Garling and Tassava, 1984).

Except for the unpublished observations of Popiela (1973) on denervated limb stumps, electron microscopy studies of dedifferentiation have been restricted to innervated limb stumps (Hay, 1966b; Salpeter and Singer, 1960b; Norman and Schmidt, 1967). Popiela concluded that the cells that appear in denervated limb stumps are not typically dedifferentiated cells; nor have biochemical studies conclusively shown qualitative differences between the RNAs and proteins of the two limb types, even though modern technology might now make differences resolvable. Data from one-dimensional polyacrylamide gel electrophoresis have so far shown few differences in stainable banding patterns (Dearlove and Stocum, 1974) and no differences in [^{35}S]methionine-labeled bands (Garling and Tassava, 1984).

Whether a wound epidermis is or is not functional on a denervated limb stump is difficult to test without a biochemical "marker" of wound epidermis functions. Cell division is so limited in a denervated limb that a cell cycle promoting role of the wound epidermis would be ineffective (Tassava and Mescher, 1975).

It is clear that the LI does increase in denervated limb stumps (Tassava et al., 1974; Mescher and Tassava, 1975; Maden, 1978), and microspectrophotometry has shown that the [^{3}H]thymidine incorporation represents nuclear DNA

TABLE 5.2. Molecular and Cellular Events Compared in Innervated and Denervated Limb Stumps of Urodeles During the Early Initiatory Days of Regeneration

Parameter	Limb type: Innervated	Limb type: Denervated	References
Wound epidermis	Yes	Yes	Kelly and Tassava, 1973; Mescher and Tassava, 1975
Dedifferentiation	Yes	Yes	Butler and Schotté, 1941; Kelly and Tassava, 1973; Mescher and Tassava, 1975
RNA synthesis	Yes	Yes	Kelly and Tassava, 1973
Protein synthesis	Yes	Yes	Garling and Tassava, 1984; Dearlove and Stocum, 1974
DNA synthesis	Yes	Minimal	Mescher and Tassava, 1975; Maden, 1978
Mitosis	Yes	Minimal	Kelly and Tassava, 1973; Mescher and Tassava, 1975; Maden 1978
Blastema	Yes	No	Butler and Schotté, 1941; Kelly and Tassava, 1973

replication (Mescher and Tassava, 1975). The LI drops off to near zero when larvae or adult newts are given [^{3}H]T after 1 or 2 weeks postamputation, respectively. Because the MI was so low, Tassava and Mescher (1975) hypothesized that the cells of denervated limb stumps blocked *after* DNA replication but *before* mitosis and that nerves may therefore regulate one or more G_2 events (see Fig. 5.2). That nerves act in the G_1 phase (Maden, 1978; Olsen and Tassava, 1984) or throughout the cell cycle (Globus, 1978) has also been suggested. In the experiments of Kelly and Tassava (1973), Mescher and Tassava (1975), and Maden (1978), a slight increase in the MI was observed in denervated limb stumps. Many questions remain. Do all the cells that replicate DNA in denervated limb stumps complete mitosis? If they complete mitosis (M phase), do they stop in G_1 or continue through another round of DNA replication and block in their second G_2? Do some cells block in their first G_1, that is, immediately after dedifferentiation? During the past few years, we have initiated numerous experiments designed to answer these and related questions.

We knew from the early work of Butler and Schotté (1941) that denervated, amputated limbs of *Ambystoma* larvae will regenerate upon reinnervation. We found that the MI increased as nerves reached the distal limb tip (Petrosky et al., 1980; Olsen and Tassava, 1984). More recently, we have shown that at least some of the cells that replicate DNA on day 4 postamputation are still in the limb stump on day 13 (the time of reinnervation), can be rescued by nerves, and contribute to the reinnervation blastema (Olsen and Tassava, 1984; Olsen et al., 1984). The experimental design of this study is shown in Figure 5.3. The observation that cells in denervated limb stumps can be rescued by nerves has positive implications for the establishment of a bioassay for the NTF based on the in vivo stimulation of cell cycling. It has been suggested that

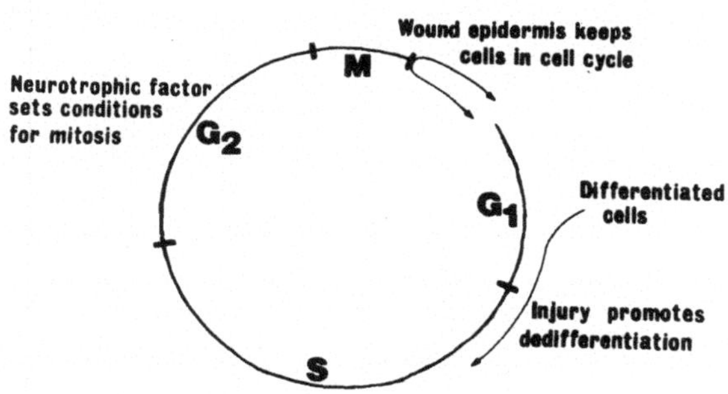

FIG. 5.2 A hypothesis explaining the roles of injury, nerves, and the wound epidermis in the initiation of limb regeneration. The hypothesis presents a generalized cell cycle of a dedifferentiating cell of the limb stump. (From Tassava and Mescher, 1975, reprinted by permission.)

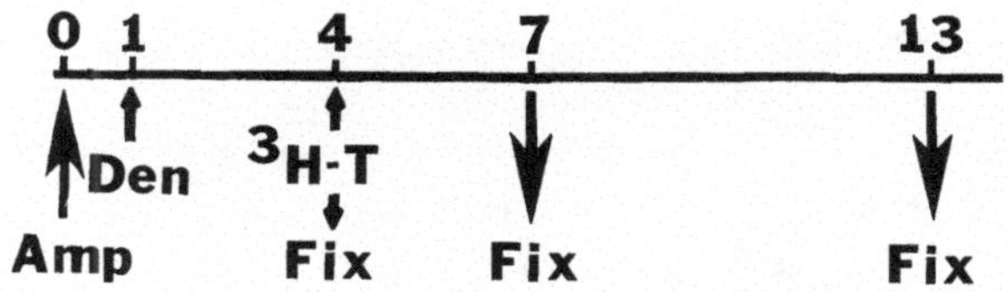

FIG. 5.3 Protocol of the "rescue" experiment, showing the days on which experimental manipulation and fixation were performed. Numbers above the line represent days after amputation, which was on day 0 (Den, denervation of left forelimb; ^{3}H-T, intraperitoneal injection of [^{3}H]thymidine; and Fix, fixation of limbs in Bouin's fixative).

cells in denervated limb stumps are rescued from the G_1 phase of the cell cycle by ingrowing nerves (Olsen and Tassava, 1984; Olsen et al., 1984).

Maden (1978) and Mescher (1984) showed that postamputation [^{3}H]T labeling indices and mitotic indices were lower in axolotl and newt limb stumps, respectively, which had been predenervated compared to limb stumps denervated on the day of amputation, and suggested that predenervation provided time for residual neurotrophic factor to be degraded. However, because predenervation resulted in a reduction in the LI but not a complete absence of labeling, we have recently considered the hypothesis that NTF can be depleted from limbs only when cells are proliferating. If this hypothesis is true, no amount of predenervation time will completely deplete a limb of residual NTF if the limb is not amputated. Only if an injury causes the cells to cycle (Tassava and Mescher, 1975) will NTF be used up. One test of this hypothesis would be to examine cellular events in limb stumps after long predenervation periods (Liversage and McLaughlin, 1983; Scadding, 1984; Mescher, 1984).

The view that NTF can be retained in limbs after denervation, as residual neurotrophic activity, is shown diagrammatically in Figure 5.4. The following question can be raised: If a limb having no residual trophic activity is amputated, will any cell cycling activity occur? Figure 5.5 shows an experimental design that we used in an attempt to answer this question (Olsen, 1980, 1982). Axolotl blastemas were allowed to develop to the notch stage (Stocum, 1979), at which time the MI was still quite high, and denervation was then performed. The cells continued to proliferate (presumably until NTF was used up) and the blastemas developed into small limbs, theoretically depleted of all NTF. These depleted regenerates were then amputated and redenervated and MI and LI were examined for the subsequent week. The LI and MI still showed small increases and therefore it can be tentatively concluded that NTF is required for even the first cell cycle of some but not all dedifferentiating cells.

Residual NTF may be influential in the differing responses of nerve-dependent and nerve-independent blastemas to denervation (see above). When nerve-dependent blastemas of axolotls were denervated, the MI, which

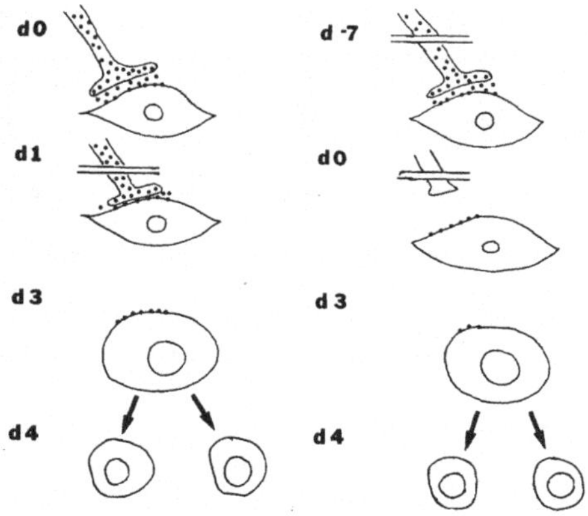

FIG. 5.4. A diagrammatic view showing how residual neurotrophic factor might persist in a limb and why predenervation does not entirely abolish cell cycling upon amputation: (left) denervation at time of amputation and (right) denervation 7 days prior to amputation.

was 2.0% at the time of denervation, continued to increase for 3 days in parallel with the MI of control, innervated blastemas. Finally, on the 4th day after denervation, the MI of denervated blastemas diverged from the controls and then dropped to zero by 7 days (Tassava and McCullough, 1978). It may be significant that 4 days is slightly longer than 2 cell cycles (McCullough and Tassava, 1976). Singer and Craven (1948) obtained the same result with nerve-dependent blastemas of adult newts. At the time the MI diverges from that of controls, it is also important to note that nerve-dependent blastemas begin to

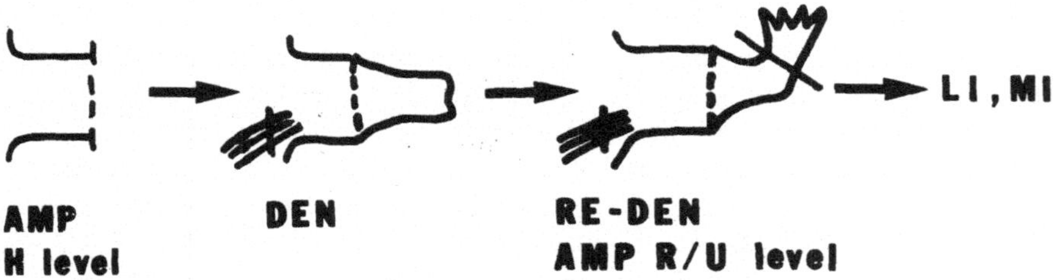

FIG. 5.5. Experimental protocol to entirely deplete limbs of neurotrophic factor prior to amputation. Limbs of newts or *Ambystoma* larvae are amputated (AMP) at the humerus (H) level. When blastemas reach a nerve-independent stage (e.g., notch), limbs are denervated (DEN). The cells that continue to cycle deplete the regenerate of neurotrophic factor. When full differentiation has occurred, the limb is reamputated through the R/U (radius and ulna) and redenervated. The LI and MI are then examined through the first week postamputation. (After Olsen, 1982.)

resorb (Schotté and Butler, 1944; Tassava and McCullough, 1978). When nerve-independent blastemas are denervated, the effect on mitosis is not nearly as adverse, and the blastemas develop into limbs morphogenetically normal but reduced in volume (Singer and Craven, 1948).

As pointed out above, the MI is adversely affected when nerve-dependent blastemas are denervated. One can ask: When cells stop cycling, in which phase of the cell cycle do they block? According to preliminary results of Tassava and McCullough (1978), the MI drops before the LI. Thus cells are replicating DNA but not undergoing mitosis, a result consistent with a G_2 block point. Attempts to detect an accumulation of G_2 versus G_1 cells in denervated blastemas by microspectrophotometry have been inconclusive with regard to resolving the block point. In one such study, Maden (1979a) denervated cone stage axolotl blastemas and found a postdenervation accumulation of G_1 cells, that is, cells with the 2C DNA content. However, cone stage blastemas are nerve independent (Schotté and Butler, 1944) and would be expected to continue developing after denervation, with the cells predictably differentiating from the G_1 phase, that is, with the 2C DNA content. In a somewhat similar study, the nuclear DNA content of both ND and NI adult newt blastemas was measured after denervation (Loyd and Connelly, 1981). As predicted, cells accumulated in G_1 in nerve-independent blastemas, but in neither case was an accumulation of G_2 (4C DNA) cells seen. The possibility exists that G_2 and/or G_1 blocked cells may be removed from the blastemas during resorption or, because of poor staining and/or abnormal histological appearance, may inadvertently be excluded from the sample.

The exact mechanism by which nerves influence cell proliferation during limb regeneration is not yet understood. Instead of comprehensively reviewing all of the cell cycle work that has been done on regeneration, we hope to have defined the problem areas and provided a theoretical framework for additional investigations.

Summary

An essential feature of salamander limb regeneration is the formation of a blastema at the end of the limb stump. Because a blastema does not form when an amputated limb is denervated, much research has been designed to elucidate the role of nerves in regeneration. Comparisons of the levels of mitotic and [^{3}H] thymidine labeling indices of innervated and denervated limb stumps or blastemas have provided results consistent with the following hypothesis: nerves provide a chemical (the so-called neurotrophic factor) necessary for the sustained cell cycling which leads to blastema formation and growth.

From available data it is not possible to distinguish between two suggested mechanisms by which the nerve influence is provided to cells of the blastema. The neurotrophic factor may be freely diffusible and potentially available to all blastema cells or, alternatively, blastema cells may be stimulated to cycle only when direct or close contact with neurites is made.

Evidence is presented for nerves acting in both the G_1 and G_2 phase of the cell cycle, but most of the data supports a major G_1 role. Data from two experimental systems utilizing larval *Ambystoma* provide important further insights into the cell cycle role of nerves. First, cells of denervated limb stumps block in the G_1 phase of the cell cycle and can be "rescued" by ingrowing nerves. This system may lead to an in vivo bioassay for the neurotrophic factor. Second, after amputation of limb stumps or regenerates previously depleted of neurotrophic factor, only a low level of cell cycling is initiated. Therefore, without neurotrophic factor, some cells are capable of at least one round of the cell cycle, while other cells cannot progress through their first G_1.

Acknowledgments

We are grateful for research support from the National Science Foundation (Grants PCM-8103430 and PCM-8315428) and to Dr. A. L. Mescher for suggestions on the manuscript. Cherie L. Olsen was a National Science Foundation predoctoral fellow.

6 Trophic Factors from Nerves

ROBERT L. CARLONE AND ANTHONY L. MESCHER

Neurons interact with target cells, including other neurons, to regulate their structure, function, growth, and differentiation (Guth, 1968; Drachman, 1974; Smith and Kreutzberg, 1976; Singer, 1978). These neurotrophic[1] interactions typically have time courses much longer than those associated with synaptic transmission and alterations in membrane potentials (Drachman, 1974). Trophic interactions are phylogenetically ancient and ubiquitous in the animal kingdom. For example, the "primitive" nervous system of *Hydra*, composed of a nerve net of multifunctional cells, directs the polarity of regeneration in that animal (Schaller, 1979). In mammals, nerves are responsible for maintenance of the structural and functional integrity of many tissues, including taste buds and skeletal muscles (Guth, 1968, 1971).

The precise mechanisms through which neurotrophic influences are mediated are unknown. Typically, nerves control a number of properties of their target cells that may be regulated by an array of mechanisms, possibly varying from tissue to tissue or from species to species. Drachman (1974) has stressed the importance of quantifying the contribution of any one mechanism to a particular trophically modulated phenomenon and suggested several possible mechanisms to explain the underlying cellular interactions. These include the release of neurotransmitters from nerve endings, activity of the target cell or tissue (e.g., as in the contraction of skeletal muscle), depolarization of the target cell membranes, physical contact between neurons and target cells, and the release of trophic chemicals distinct from the natural neurotransmitters. The abundant variety of proposed neurotrophic mechanisms need not be mutually exclusive and it is possible that two or more could be involved in any one trophic interaction.

One phenomenon in which the trophic effects of nerves on their target tissues are reasonably well defined and that apparently involves the synthesis, transport, and release of distinct neurotrophic factors is the stimulation of growth of the amphibian limb blastema after amputation (Singer, 1965, 1974, 1978; Wallace, 1981; Brockes, 1984).

Nature of the Neurotrophic Influence on Amphibian Limb Regeneration

The general neural control of the early events of regeneration has been reviewed in Chapter 5 and therefore needs only a brief reintroduction here. Intact nerves are essential for the early proliferative phase of regeneration.

Generally, denervation of the limb by surgical resection of brachial nerves 3, 4, and 5 at any time prior to onset of the differentiative phase (or palette stage; see Iten and Bryant, 1973) results in a failure of the limb to regenerate, provided that reinnervation is prevented (Schotté, 1926b). In contrast, denervation after the onset of this stage fails to inhibit further redifferentiation and morphogenesis of the blastemal tissues into an intact, functional limb (Schotté, 1926b).

At the cellular and molecular levels, denervation results in reduced rates of DNA, RNA, and protein synthesis (Dresden, 1969; Lebowitz and Singer, 1970; Singer and Caston, 1972) and mitosis (Singer and Craven, 1948) in the blastemal tissues. It can thus be argued that the basic function of nerves in this regeneration process is to enhance growth of the early blastema via a stimulatory effect on the rates of macromolecular synthesis and proliferation of the blastemal cells. This trophic role is independent of the type of nerve supplying the regenerate. It has been shown that motor nerves (Sidman and Singer, 1960), sensory nerves (Singer, 1943, 1945), or central nervous system tissue alone (Singer, 1952) can stimulate blastemal growth, provided that an adequate supply of intact axons is present at the amputation surface (Singer, 1965; Rzehak and Singer, 1966).

What, then, is the precise nature of the nerve's growth stimulus? Since it is apparently common to both sensory and motor nerves, electrical impulse transmission is unlikely to be involved. Drachman and Singer (1971) have demonstrated that normal regeneration can occur in limbs during prolonged paralysis induced by botulinum toxin, which blocks acetylcholine release. Moreover, acetylcholine,when injected into denervated urodele limbs (Singer, 1960) or added to cultured limb regenerates (Foret and Babich, 1973a), is ineffective as a growth-promoting substance.

Other neurotransmitters have been proposed as the trophic agent. Schotté (1926b) first championed the cause of norepinephrine, later to be supported by Taban et al. (1978). Norepinephrine, however, unlike brain extracts, is ineffective in maintaining the rates of macromolecular synthesis and mitosis in denervated and cultured blastemas (Rathbone et al., 1980). In addition, it was found that limbs regenerate normally in the absence of sympathetic innervation (Singer, 1942). A central role for norepinephrine in the growth stimulus is therefore suspect.

A recent preliminary report that the undecapeptide neurotransmitter, substance P, has some mitogenic activity when added to blastemal cultures (Globus et al., 1983) is intriguing, given the distribution of this peptide in the nervous system of vertebrates. Substance P is relatively abundant in the hypothalamus, sensory nerves, and both afferent and efferent sympathetic fibers (Jessel, 1981), but has not been shown to be present within motor neurons, which alone, in sufficient density, can support regeneration (Singer, 1945; Sidman and Singer, 1960). Further information on the possible trophic role of this neurotransmitter in the regeneration process is needed.

As previously mentioned, sensory nerves alone can support regeneration (Singer, 1943, 1945). Therefore use or repeated activation of the target tissue is not likely to be a contributing factor to the nerve's trophic maintenance in this system, especially in light of the regeneration observed by Drachman and Singer (1971) in amputated limbs treated with botulinum toxin.

Physical contact may not be central to the growth stimulus, despite studies that demonstrate a close association between growing neurites and blastemal cells, even occasionally involving desmosomal contacts, during the nerve-dependent phase of regeneration (Salpeter, 1965; Schmidt, 1968). For instance, infusion of aqueous extracts of spinal ganglia (Lebowitz and Singer, 1970) or brain (Deck, 1971; Singer et al., 1976) into denervated adult newt limb regenerates mimics at least some effects of innervation, particularly on the rates of macromolecular synthesis in blastemal cells. In addition, mitosis can be promoted in cultured blastemas by the addition of brain extracts (Carlone and Foret, 1979) or by sensory ganglia cultured transfilter to blastemas on Millipore filters of extremely low porosity (0.05 μm) (Tomlinson et al., 1981).

We are thus left with the possibility proposed by Singer (1960, 1965) that trophic chemicals distinct from the known neurotransmitters are responsible for the growth stimulus of nerves during limb regeneration in amphibians. Such chemicals are probably synthesized in the neuronal perikarya, transported down the axons, and released at the end organs (Van Arsdall and Lentz, 1968; Singer, 1974). Indeed, evidence exists for the transport of [^{3}H]leucine-labeled proteins during axonal regeneration in both larval (Scadding and Liversage, 1979) and adult (Lentz, 1972) limbs. Wallace (1981) has utilized length estimates of distal portions of cut axons coupled with temporal measurements of decreases in the rates of macromolecular syntheses in denervated blastemas (Singer and Caston, 1972) to implicate the fast component of axoplasmic transport (100–400 mm/day; see Ochs, 1974) in the delivery of trophic factors to the target tissue. Wallace (1981) further maintains that materials transported by fast axoplasmic flow might include newly synthesized proteins such as the cholinesterases, as well as chemicals like phosphatidylcholine and most transmitter substances (Ochs, 1974). Since many of these have previously been eliminated as candidates (Singer, 1960, 1974), the identity of the neurotrophic factor or factors remains to be determined.

Attempts to Isolate the Neurotrophic Factor(s) for Limb Regeneration

Most attempts to isolate a distinct neurotrophic factor(s) have used a replacement strategy whereby substances or extracts are either injected or infused into denervated limbs or added to the culture medium of explanted blastemas to determine whether they (1) support limb regeneration in the absence of nerves or (2) prevent denervation-induced changes in the rates of macromolecular synthesis and/or mitosis. Efforts to stimulate complete regeneration in denervated limbs by such means have proven unsuccessful (Kamrin and Singer, 1959; Singer, 1960; Deck and Futch, 1969; Deck, 1971), although a positive effect on growth was observed in some cases (Kamrin and Singer, 1959; Burnett et al., 1971). Growth stimulation reported with blastemal extracts by Burnett et al. (1971) is, however, likely due to reinnervation of the distal limb stump.

In Vivo Bioassays

A more precise and sensitive assay of the nerves' trophic influence on blastema growth is the ability to prevent decreases in the rate of protein synthesis characteristic of denervation (Dresden, 1969; Lebowitz and Singer, 1970). Lebowitz and Singer (1970) tested the ability of crude homogenates and clarified supernatants of newt brachial nerves to stimulate [^{14}C]leucine incorporation into soluble blastemal proteins when infused for 1 hr into denervated adult newt limbs. The homogenates abolished approximately 60% of the denervation-induced inhibition of protein synthesis. The trophic activity of the homogenates was stable at −20° C for 24 hr, but was destroyed by heating at 90° C for 10 min, firmly establishing the chemical nature of the neurotrophic influence.

The pioneering studies of Lebowitz and Singer (1970) were followed by further attempts by Singer and his colleagues to isolate and characterize the trophic factor(s) (Singer et al., 1976; Jabaily and Singer, 1978). Using the same infusion bioassay, Singer et al. (1976) demonstrated that the trophic activity was not unique to peripheral nerves, but was also present in 140,000 × *g* supernatants of newt brain homogenates. Extracts prepared in a similar manner from liver were only slightly effective, and those from spleen ineffective, in stimulating protein synthesis. The neurotrophic activity was heat labile (60° C for 10 min), trypsin sensitive, probably basic, and could be stored frozen for at least 2 months with little loss of potency. Based on centrifugation characteristics, the molecular weight of the brain trophic activity was estimated to be less than 300,000.

Synaptosomal preparations from frog brain were equally effective in stimulating protein synthesis in denervated blastemas, indicating (1) a lack of species specificity of the trophic agent and (2) its localization within the neuronal fraction (and specifically, axon terminals) of nervous tissue (Singer et al., 1976).

In Vitro Bioassays

Further efforts to purify a trophic activity from brain tissue have necessitated the development of a more rapid and reliable bioassay. Although the slow infusion methods of Deck (1971) and Singer et al. (1976) probably most closely resemble the postulated delivery of trophic factors to the blastema, they are time-consuming, technically demanding, and not without limitations. The successful explantation and maintenance of amphibian blastemal tissues in organ culture (Stocum, 1968a; Babich and Foret, 1973; Foret and Babich, 1973a,b; Globus and Liversage, 1975b) has led to the development of a number of in vitro bioassays for the trophic factor (Globus and Vethamany-Globus, 1977; Choo et al., 1978; Carlone and Foret, 1979) that avoid such infusion-related problems. These in vitro bioassays, based on the ability of substances added to the culture medium to reverse the explantation-induced decreases in the rates of protein and DNA synthesis (Choo et al., 1978; Vethamany-Globus et al., 1978; Mescher and Loh, 1981) or mitosis (Globus and Vethamany-Globus, 1977; Carlone and Foret, 1979), have their own limitations, as do any in vitro studies. For example, interpretation of experimental results in the absence of possible contributions by unknown physiological factors make their extrapolation to the in vivo situation difficult. Nevertheless, Choo et al. (1978) and Mescher and Loh (1981) have demonstrated that the alterations in the rates of macromolecular synthesis in cultured regenerates are quantitatively and qualitatively (but not temporally) similar to those observed in denervated blastemas in vivo (Singer, 1974) and can be reversed by the addition of both newt and embryonic chick brain extracts. The trophic activity in embryonic chick brain 105,000 $\times g$ supernatants is heat labile, trypsin and protease sensitive, and RNAse insensitive. Unlike Singer's factor(s) from newt brain (Singer et al., 1976), the chick brain extract's effectiveness diminished rapidly when stored frozen (Choo et al., 1981).

Choo et al. (1981) reported further progress in their attempts to purify and characterize the trophic activity in chick brain extracts. DEAE-cellulose ion-exchange chromatography of the crude supernatant produced a single region of activity that bound only loosely, or not at all, to the anion exchanger and was thus probably composed largely of basic proteins (pI $>$ 7.2; see Choo et al., 1981), confirming earlier reports by Singer et al. (1976) with newt brain homogenates. Alternative fractionation of the crude supernate by polyacrylamide gel electrophoresis resolved the trophic activity into two components: F (fast running) and S (slow running). Additional electrophoresis in the second dimension in the presence of SDS (sodium dodecyl sulfate) revealed at least five stainable protein spots common to the two regions of activity in both 7.5 and 10% native acrylamide gels. These proteins had apparent molecular weights of $F_1 = 13{,}500$, $S_1 = 47{,}000$, $S_2 = 36{,}000$, $S_3 = 28{,}000$, and $S_4 = 13{,}500$, respectively. Unfortunately, it was impossible to determine which (if any) of the five stainable protein spots present in either region were active in the blastemal bioassay. The possibility of multiple factors contributing synergistically to the overall growth stimulus was raised by the authors.

Our most recent results (Carlone and Rathbone, 1985) tend to support the concept of a multifactorial growth stimulus for blastema cells in culture. Using a bioassay for neurotrophic activity similar to that of Mescher and Loh (1981), we have partially purified a low molecular weight, trypsin-sensitive, and heat-stable factor from adult chicken brain. This chicken brain growth factor (CBGF) is a potent mitogen at nanomolar concentrations in vitro for both newt limb blastema cells and Swiss mouse 3T3 fibroblasts.

The CBGF has a pI greater than 8.4, as revealed by chromatofocusing, and an apparent molecular weight of 2000–6000. This mitogenic activity is specific to extracts of neural tissue but lacks species specificity, being detectable in aqueous extracts of newt and both embryonic (10 day) and adult chicken brain.

While many of these characteristics are consistent with results of the previously mentioned attempts to purify neurotrophic factors for blastemal growth from brain tissue, others are unique. Chicken brain growth factor is considerably smaller than the factors of Singer et al. (1976) and Choo et al. (1981) and, unlike either, extremely heat stable. Moreover, CBGF is electrophoretically and chemically distinct from other low molecular weight peptides, such as substance P and epidermal growth factor, with demonstrated neurotrophic activity on cultured blastemas (Mescher and Loh, 1981; Globus et al. 1983). The relationship of CBGF to other brain-derived, low molecular weight peptide mitogens is presently unclear.

Our results and those of Choo et al. (1981) are, however, consistent with Singer's proposed model for trophic communication between cells (Singer, 1978; Jabaily and Singer, 1978), involving a family of polypeptide factors as universal mediators of the rates of synthetic and replicative events in target cells. Insights into the mechanism underlying the neurotrophic influence on blastemal cells may be derived from a closer examination of various growth factors and maintenance factors that have recently been isolated from neural tissues using other bioassays.

Growth Factors from Neural Tissues

Nearly fifty years ago studies on the growth of cells in vitro revealed that extracts of brain are much richer than extracts of any other adult tissue in their content of growth-stimulating factors (Trowell and Willmer, 1939; Hoffman, 1940). Since that time, and especially in the last decade, cultured cells have been widely employed to assay for neural factors with mitogenic activity. Because muscle development and survival are dependent on a trophic interaction with nerves (Guth, 1968; Gutmann, 1976), assays involving myoblast growth and differentiation in vitro are commonly used to identify neurotrophic factors. Conditions affecting the development of myoblasts in culture can be compared to the requirements for the maintenance of muscle in vitro, which is often studied by measurement of changes in acetylcholinesterase activity. Bioassays involving muscle and other cells have led, in recent years, to the isolation from mammalian or avian nervous tissue of several factors with "trophic" activities defined according to the system by which they are measured. In no case is the actual physiological or developmental role of the factor known with certainty, and it is possible that proteins similar to one or more of these may be involved in the neurotrophic effect seen during limb regeneration.

Fibroblast Growth Factor

The first growth-promoting agent to be purified from nervous tissue was fibroblast growth factor (FGF), isolated by Gospodarowicz (1974) from protein fractions of bovine brain and pituitary, using a line of fibroblasts to assay for mitogenic activity. Characterized as a basic polypeptide with an apparent molecular weight of 13,000, FGF was found in early studies to promote the growth in vitro of a wide variety of cells, especially cells derived from the embryonic mesoderm (Gospodarowicz et al., 1978b). Mammalian cells of connective tissue origin (fibroblasts, chondrocytes) responded to FGF at concentrations as low as 0.1 ng/ml of medium, and mammalian myoblasts showed optimal growth and development in medium supplemented only with horse serum (10%) and FGF at 100 ng/ml (Gospodarowicz and Mescher, 1977). Fibroblast growth factor was also shown to maintain total acetylcholinesterase activity of newt triceps muscle in organ culture (Carlone et al., 1981). When tested in vitro on early regeneration blastemas from newt forelimbs, FGF was found to stimulate DNA synthesis (Mescher and Loh, 1981) and mitosis (Carlone et al., 1981) in a dose-dependent manner. Fibroblast growth factor caused renewed mitotic activity in mesenchymal cells when infused into denervated blastemas in vivo (Mescher and Gospodarowicz, 1979), although sustained growth was not achieved by such an approach. Gospodarowicz et al. (1975) also reported that local administration of FGF to normally nonregenerating amputated forelimbs of adult frogs (*Rana pipiens*) produced a regeneration response comparable to that found by Singer (1954) after surgical augmentation of the nerve supply to the forelimbs of much smaller frogs.

Studies on the chemical properties of brain FGF have been contradictory. The basic nature of the polypeptide has been demonstrated repeatedly by Gospodarowicz and co-workers (1978a, 1982, 1984). Immunological and peptide mapping studies (Westall et al., 1978) indicated differences between brain FGF and pituitary FGF and identified the former as a fragment of myelin basic protein (MBP) generated by an endogenous protease. The suggestion that MBP may serve as a precursor for FGF was challenged by workers who acknowledged that MBP fragments represented more than 98% of the protein in the "purified" FGF, but asserted that the active mitogen in the preparation is a different, more acidic polypeptide (Thomas et al., 1980, 1984). An acidic growth factor with a molecular weight of 16,600 has recently been purified from bovine brain by Thomas et al. (1984). Gospodarowicz et al. (1982) agreed that acidic growth factors are present in brain tissue, but their isoelectric focusing studies showed nearly all of the mitogenic activity focusing in the basic range (pI 9.6). The most recent work of Gospodarowicz et al. (1984), using a novel method of purification, indicates that the basic mitogen is similar in both brain and pituitary and, as suggested by Thomas et al. (1980), is unrelated to MBP, the fragments of which could not be separated from FGF in other purification procedures.

It should be noted that at least two other groups of investigators have isolated from nervous tissue growth factors with chemical and biological properties similar to those originally reported for FGF. Jennings et al. (1979) purified from bovine spinal cord a mitogenic polypeptide with a molecular weight of 11,000 and a pI of 9.5–9.8. As mentioned in the previous section, basic polypeptides that stimulate protein synthesis in cultured regeneration blastemas have also been partially purified from embryonic chick brain by Choo et al. (1981). The relationship of these proteins to the factor of Gospodarowicz et al. (1984) remains to be determined.

Owing partly to the lack of information regarding its structure, the mechanism of action of FGF has not been as extensively studied as those of certain other growth factors. Evidence is increasing, however, that at least part of its mitogenic effect may be indirect. Gospodarowicz (1983) has reviewed work showing that several cell types respond to growth-promoting agents in vitro only in the presence of a well-developed extracellular matrix (ECM) and that FGF induces the synthesis and secretion of various ECM components by cultured cells. The ECM is of critical importance to many fundamental cellular activities, such as cell shape changes, migration, and proliferation, but the means by which it exerts its effects are poorly understood (Hay, 1981). Gospodarowicz et al. (1981) speculate that one way the ECM might affect cell proliferation is by modulating changes in cell shape, since these are a basic feature of cell division. In vitro studies examining both DNA synthesis and mitosis in normal embryonic cells implicate the ECM in the control of cell cycle events: low-density cell cultures maintained on an ECM show a high DNA labeling index and a high mitotic index, while similar cultures maintained on a suboptimal substrate (bare plastic) have a high DNA labeling index but a low mitotic index, suggesting a failure of the cells to enter into mitosis (Yaoi and Kanaseki, 1972). Such a failure might be related to an inability of cells on an inappropriate substrate to undergo at a normal rate the characteristic change in morphology (rounding up) that precedes cytokinesis. The fact that similar cell cycle data, that is, normal DNA labeling indices but low mitotic indices, have been found in the dedifferentiating cells of amputated–denervated urodele limbs (Tassava and Mescher, 1975; see also Chap. 5) indicates that further investigation of a possible functional relationship between the ECM and neurotrophic factors during regeneration is warranted.

Sciatin/Neurotransferrin

Another trophic factor, purified from chicken sciatic nerves by Markelonis et al. (1980) and named sciatin, has been studied primarily with regard to its effects on myogenesis in vitro. Sciatin has been shown to (1) promote the fusion of embryonic chick myoblasts into well-developed, multinucleated myotubes (Oh and Markelonis, 1980), (2) stimulate protein synthesis and creatine kinase activity in embryonic myoblast cultures (Markelonis and Oh, 1978), and (3) maintain the level of acetylcholinesterase in such cultures and in organ cultures of adult muscle (Oh and Markelonis, 1978). Specific localization of sciatin in neuronal cell bodies and axons has been demonstrated immunocytochemically (Oh et al., 1981). The importance of this factor in interpreting results of previous studies of muscle development in vitro is underscored by the finding (Oh and Markelonis, 1980) that immunoabsorption with antiserum against sciatin removed all myotrophic activity from chick embryo extract, for many years a routine supplement for myoblast culture media.

Sciatin is an acidic glycoprotein with a molecular weight of approximately 80,000 (Markelonis et al., 1980). Amino acid sequence analysis and immunological studies (Markelonis et al., 1982) indicated that sciatin is structurally related to transferrin, the major iron transport protein in vertebrate blood. Chicken serum transferrin was subsequently shown to duplicate the biological effects of sciatin on myoblasts (Oh and Markelonis, 1982). A factor similar both chemically and biologically to sciatin/transferrin was independently isolated from chicken ischiatic–peroneal nerves by Popiela and Ellis (1981), who introduced the name neurotransferrin (Popiela et al., 1984). Other investigators have purified a factor with sciatin-like properties from chicken serum and from chick embryo extract and identified it as a transferrin (Kimura et al., 1981; Ii et al., 1981, 1982). Besides confirming the results of Oh and Markelonis (1980) with regard to the importance of transferrin in embryo extract, these workers also demonstrated that proper concentrations of ferric salts alone are capable of exerting myotrophic effects (Ii et al., 1982; Hasegawa et al., 1981). Whether neurotransferrin/sciatin purified from peripheral nerves is taken up from intercellular fluid by neurons or synthesized within the perikarya is not yet known. Support for the latter possibility is provided by the recent demonstration that cultured chick embryo spinal cord neurons synthesize a transferrin-like protein with myotrophic activity (Stamatos et al., 1983).

The identification of transferrin as a component of the trophic activity of neurons is especially interesting in light of the fact that this protein is a well-known requirement for the proliferation of cells in vitro (Barnes and Sato, 1980). Since iron is a necessary cofactor in key enzymes involved in energy metabolism and DNA synthesis (Bezkorovainy, 1980), the growth-promoting effect of transferrin apparently stems from its receptor-mediated ability to deliver iron to cells in an appropriate manner (Morgan, 1981). As a supplement to defined (serum-free) medium, transferrin is routinely added at doses ranging from 0.5 to 50 μg/ml (Barnes and Sato, 1980), the range of concentrations also required for the effects of sciatin/neurotransferrin on myoblasts (Oh and Markelonis, 1980; Popiela et al., 1982). Mitogenic growth factors typically exert their maximal effect at ng/ml concentrations or less (Gospodarowicz and Moran, 1976).

With regard to limb regeneration, heterologous serum transferrin has been shown to stimulate DNA labeling and mitotic indices in denervated newt forelimb blastemas in vivo (A. L. Mescher, unpublished data) and in vitro (Mescher and Munaim, 1984). Measurements of [^{3}H]thymidine incorporation in the blastema culture system indicate that the transferrin effect is dose dependent and involves the delivery of iron to the cells (Mescher and Munaim, 1984). These results argue for the presence of receptors for transferrin on blastemal cells and imply a role for neuronally released transferrin in the promotion of blastemal growth, as suggested by the aforementioned work with sciatin/neurotransferrin. Transferrin from nerves, as a supplement to that available from the plasma, may thus be required for active cell proliferation in the early limb regeneration blastema, since this tissue is relatively avascular (Schmidt, 1968; Peadon and Singer, 1966; Smith and Wolpert, 1975). This proposed role for transferrin during the nerve-dependent phase of limb regeneration comes very close to the model of Singer (1978) in which a trophic factor that is ubiquitous, but delivered in greater quantities by neurons, promotes blastemal growth primarily by influencing the rate of macromolecular synthesis.

Neuronal Surface Mitogen(s) from Schwann Cells

In addition to the nerve-dependent developing systems already mentioned, neurons are known to exert a trophic influence upon Schwann cells investing peripheral axonal processes (Asbury, 1975). In recent years, cultured Schwann cells have been used to study the nature of this neurotrophic influence. Salzer and Bunge (1980) have demonstrated that Schwann cell proliferation in vitro occurs under the same two circumstances that cause such growth in vivo: upon contact with growing neurites and during nerve degeneration after injury. These authors presented evidence that the mitogenic stimuli during development and during degeneration are distinct. Their results suggest that proliferation following nerve injury is promoted by products released from the degenerating myelin, a possibility consistent with numerous in vivo observations (Abercrombie, 1957) and with recent work on the mitogenicity of degenerating myelinated and unmyelinated nerve segments and nerve extracts (Yachnis and Mescher, 1982; Mescher, 1983). The trophic stimulus during development of peripheral nerves, however, is apparently related to a trypsin-sensitive factor located on the neurite surface, since a membrane fraction prepared from cultured neurites (Salzer et al., 1980a) or from brain (DeVries et al., 1983), but not membrane preparations from other cells, is mitogenic for cultured Schwann cells. No trophic activity in this bioassay was associated with the cytosol-enriched fractions (Salzer et al., 1980b). Moreover, stimulation of Schwann cell proliferation by neurons was dependent on direct contact between neurites and Schwann cells: separation of the two cells by a collagen diaphragm that allowed diffusion of relatively large proteins, but prevented neurite penetration and cell–cell contact, resulted in no Schwann cell growth (Salzer et al., 1980b).

Whether cell contact and involvement of factors in the axolemma will prove to be important in other neurotrophic phenomena remains to be determined. With regard to the role of nerves in promoting blastema formation, it is important to reiterate that although growing neurites may form close associations with blastemal cells (see above), direct contact is apparently not required for the in vitro stimulation of blastemal cell proliferation by nerves (Tomlinson et al., 1981) or neural extracts (Carlone and Foret, 1979). The recent demonstration that a Schwann cell mitogen is also present on the surface of pheochromocytoma PC12 cells (Ratner et al., 1984) may lead to the factor's purification and increased understanding of its mechanism of action.

Other Neurally Derived Trophic Factors

Several other factors with various "trophic" activities have been at least partially purified from vertebrate nervous tissue in recent years. Lim (1980) has reviewed work on glia maturation factor, a large (200,000-dalton) acidic protein prepared from mammalian brain that promotes glioblast differentiation in vitro. A basic protein with a molecular weight of 31,000 that stimulated proliferation of cultured glial and Schwann cells has been purified from bovine brain and pituitary (Lemke and Brockes, 1984). Similar activity has been detected in extracts of newt brain and blastemas (Brockes, 1984). A protein with the same pI and molecular weight has been partially purified from rat brain by Lentz et al. (1981), assaying for the ability to maintain acetylcholinesterase activity in cultured muscle. Low molecular weight peptides have been extracted from chicken brain and spinal cord that induce acetylcholine receptors on embryonic chick myotubes in vitro (Jessel et al., 1979). Proliferation of cultured endothelial cells is stimulated by an acidic protein isolated from bovine hypothalamus by Maciag et al. (1982). This activity may be identical to that purified by Thomas et al. (1984). As discussed earlier, such factors are characterized biologically (and frequently named) according to the bioassay used to detect them. Various other activities, including a nonspecific trophic or mitogenic effect, may be ascribed to these proteins in the future.

Conclusions

Neurotrophic interactions involve relatively long-term communications between nerves and their target tissues and result in the stimulation of growth or the modulation of many differentiated characteristics of the target cells. Nerves may exert their trophic influences through a variety of mechanisms. One such mechanism, involving the synthesis, transport, and release of neurotrophic factors distinct from the known neurotransmitters, has been implicated in several developmental and reparative processes throughout the animal kingdom. A prime example is the regeneration of urodele amphibian appendages. This process is normally dependent upon the presence of an adequate supply of intact nerve fibers in the limb stump whose primary contribution appears to be the stimulation of mesenchymal cell proliferation and consequent acceleration of blastemal growth. Most evidence indicates that the growth-promoting stimulus resides in an as yet unidentified protein(s). Recent results from attempts at their isolation and purification from brain tissue indirectly support this contention, as well as the proposal by Singer (1978) of a ubiquitous family of trophic peptides modulating molecular communication between cells in close proximity. As further support for this concept, a number of trophic and/or mitogenic peptides have been isolated from mammalian and avian neural tissues. Their effects on the proliferation, cellular differentiation, and structural–functional maintenance of a variety of cultured cells and tissues are becoming increasingly well characterized. With this increased understanding comes the revelation that some of these isolated peptides are biochemically similar not only to each other, but also to previously isolated proteins or glycoprotein factors with heretofore seemingly unrelated physiological roles. The possible contribution made by these factors, as well as those of various well-known hormones and peptide neurotransmitters, to the neural stimulation of blastemal cell proliferation is at present unclear and will no doubt occupy a prominent position in future work on the neurotrophic control of limb regeneration.

Note: Recent work has clarified the chemical relationships among some of the acidic mitogens discussed above. Using techniques that include heparin affinity chromatography, workers in two laboratories (Lobb, R. R., and J. W. Fett, 1984, Biochem. *23*: 6295–6299; Klagsbrun, M., and Y. Shing, 1985, Proc. Nat. Acad. Sci. U.S.A. *82:* 805–809) have shown that the acidic FGF purified from bovine brain by Thomas et al. (1984) is identical to both the endothelial cell growth factor of Maciag et al. (1982) and a bovine retina-derived growth factor described by D'Amore et al. (1981, Proc. Nat. Acad. Sci. U.S.A. *78:* 3068–3072).

With regard to neural transferrin, there is new evidence for both neuronal synthesis of this factor (Levin, M., et al., 1984, Biochem. Biophys. Res. Comm. *122:* 212–217) and its uptake by neurons from intercellular fluid (Markelonis, G., et al., 1985, J. Cell Biol. *100:* 8–17).

Acknowledgments

The authors wish to thank Debbie Pryor and Betty Monaco for typing the manuscript. The authors' research was supported in part by National Institutes of Health grants GM 27735 and GM 31080 to A. L. M. and Medical Research Council of Canada grant MA 6326 to Michel P. Rathbone and R. L. C.

Notes

1. The terms "neurotrophic" or "trophic" will be used interchangeably throughout this review to signify only those effects nerves exert on their target tissues. Target organs also influence the differentiation and outgrowth of neurons and such an effect has variously been termed "neurotrophic," "neurotropic" and "neuronotropic" in the literature, not without some confusion.

7 In Vitro Studies of Controlling Factors in Newt Limb Regeneration

MORTON GLOBUS AND SWANI VETHAMANY-GLOBUS

For the greater part of this century, the predominant experimental design utilized in regeneration studies, called for a defect approach involving extirpation of a tissue or gland, sometimes followed by replacement therapy, with a view to establishment of a set of both short- and long-range tissue interactions, essential for limb regeneration to occur. With this approach, a considerable amount of information has been accumulated over the years, demonstrating that nerves, hormones, and the apical epidermal cap are indeed the major controlling factors in limb regeneration. However, some very demanding questions remain to be answered, such as the chemical nature of trophic agents involved, as well as identification of the precise cellular and molecular events dependent upon such tissue dialogue. Recent technological advances have placed many tools in our hands, in particular in vitro methods, which are being increasingly employed as a means of understanding cause and effect relationships during growth and differentiation of regenerates. We will attempt, for the purpose of this monograph, to review what has been learned using an in vitro approach and to demonstrate some of the advantages it has to offer.

Neurotrophic Effects in Vitro

In both adult and larval urodele amphibians, the presence of an adequate supply of nerves at the wound surface is required for limb regeneration to occur (Schotté and Butler, 1941; Singer, 1952, 1974). The neural influence is essential during the early stages of limb regeneration, corresponding to the blastema accumulation and proliferation phases, and a critical mass of cells is not attained in a denervated limb (Singer and Craven, 1948). Therefore the primary role of the nerve in limb regeneration is the stimulation of mitosis.

Until recently, research on the neural control of limb regeneration was largely carried out in vivo. Although this approach has yielded and continues to generate many important contributions (see Chap. 5) our culture system (Globus, 1970; Vethamany, 1970; Globus and Vethamany-Globus, 1977) has provided a convenient means of studying nerve–blastema interactions and of pursuing the nature and mode of action of neural, hormonal, and epidermal influences. There are several important advantages to the in vitro system: (1) The regenerate and controlling factors can be isolated from the systemic circulation, thus reducing the number of unknowns in the system. (2) When a limb is surgically denervated, one must ensure that nerve fibers do not regrow into the regeneration area, and this requires one or more redenervations. Conversely, an excised blastema is effectively isolated, not only from sensory and motor neurons, but also from blood-borne factors that may influence its developmental fate. (3) In vitro, the isolated regenerate may be conveniently used to assay, for example, the activity of putative neurotrophic factors and to affirm their specific effects by using antibodies to such factors.

Evidence for a Mitogenic Neurotrophic Factor

The neural dependence of the blastema for growth and subsequent differentiation is retained in vitro and follows the normal in vivo pattern. Our early work showed that blastema cell proliferation and differentiation occur in vitro in larval *Ambystoma* tail regenerates, innervated by the intact spinal cord (Globus, 1970; Globus and Liversage, 1975a), and in adult newt limb blastema explants, innervated by implanted sensory ganglia (Globus, 1970; Globus and Liversage, 1975b); however, in these experiments ganglionic neurons were in direct physical contact with blastema cells, thereby imposing certain limitations with respect to causal analysis.

Transfilter Effects of Nerves

In order to physically isolate the regenerate from the neurons, we developed a transfilter culture system (see Fig. 7.1) in which ganglionic neurons and the blastema explant are placed on opposite sides of a thin (25-μm) Millipore filter (Globus and Vethamany-Globus, 1977). The transfilter studies provided strong evidence that the neurotrophic (mitogenic) effect is mediated by a chemical factor capable of transmission through thin filters of low porosity (0.45 μm). Of particular interest is the fact that the mitogenic effect of the neurons (which appears to be concentration dependent) is manifested in a proximodistal gradient of mitotic activity that diminishes with distance from the nerve source (Globus and Vethamany-Globus, 1977; Globus, 1978). In these initial transfilter experiments, the blastemas were cultured for 96 hr. In subsequent studies, explants grown for 48 hr exhibited a pattern of mitotic activity closely resembling 96-hr cultures. Curiously, the effect of the nerve was not manifested in an increased mitotic activity after only 24 hr of culture; instead, these explants showed a proximodistal profile similar to time-zero controls, with a characteristic peak of mitotic activity in the subapical mesenchyme.

Mitogenic Effect Retained Without Cell-to-Cell Contact Between Nerves and Blastema

Inductive interactions that occur during morphogenesis or regeneration could theoretically be mediated either by chemical exchange or by direct cytoplasmic contact between the tissues involved. Our transfilter studies (Globus and

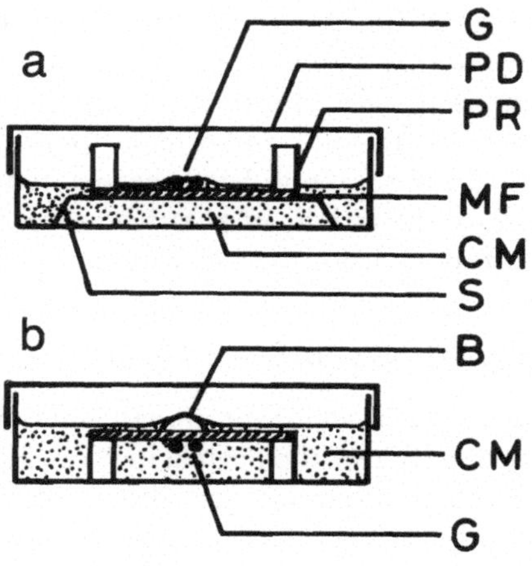

FIG. 7.1. Transfilter culture assembly. (a) Ganglia (G) cultured for at least 1 day to achieve attachment to the Millipore filter (MF) and establishment of the ganglia preparatory to introduction of the blastema (B). (b) Inversion of the filter and introduction of a blastema transfilter to the ganglia: PR, Plexiglas ring; PD, Petri dish; S, stainless steel grid; and CM, culture medium. (From Globus and Vethamany-Globus, 1977, reprinted with permission.)

Vethamany-Globus, 1977) led us to conclude that cytoplasmic processes of neither nerve nor blastema penetrate through ultrathin 0.45-μm Millipore filters to effect contact between the tissues (see also Dodson and Hay, 1974), and our more recent studies lend support to that conclusion. Firstly, when blastemas are cultured transfilter to sensory ganglia on extremely low porosity (0.05μm) filters, the mitogenic effect of the nerve is still maintained (Tomlinson et al., 1981). Secondly, the mitogenic effect of nerves is achieved when the blastema explants are cocultured 2 mm distant from the sensory ganglia, so that prevention of cell-to-cell contact is ensured (Tomlinson et al., 1981). In this case, the neurotrophic factor must be transported through the culture medium from a more distant source and in effect represents a "sustained conditioning" of the medium by neuromitogenic factor(s). Isolation of this diffusible nerve factor from the culture medium would be the next step toward identification of the neurotrophic factor(s).

Effects of Nerves on Macromolecular Events

A number of reports (Dresden, 1969; Lebowitz and Singer, 1970; Singer and Caston, 1972; Morzlock and Stocum, 1972; Kelly and Tassava, 1973; Tassava et al., 1974; Singer and Ilan, 1977) have described a post-denervation decline in the accumulation of newly synthesized macromolecules (DNA, RNA, proteins) that are paramount to cell proliferation. Our in vitro studies reveal that the incorporation of [^{3}H]thymidine into DNA (Vethamany-Globus et al.,

1978) and [^{3}H]uridine into RNA (M. Globus, S. Vethamany Globus, A. Darch, unpublished data) decreases, in the nonganglionated explants, to a level that is approximately 70% that of ganglionated blastemas. Because of this rather high residual level of macromolecular synthesis that does not appear to be subject to neural control, it is difficult to escape the conclusion that mitotic activity is a more reliable regeneration-characteristic parameter for the assay of neurotrophic activity.

Mitotic Index Profiles in Postexcision Blastemas

Although the assessment of mitotic activity is employed as a key parameter in limb regeneration studies, a considerable discrepancy can be identified in the mitotic indices reported, for example, by Litwiller (1939), Bassina (1940), Rose (1948), and Chalkley (1954). To some extent differences may be explained on the basis of blastemal stage of regeneration, but perhaps also on temporal aspects of the experiments, in particular the time of fixation after excision of the regenerated. Indeed, the number of mitoses rapidly declines as a function of time after extirpation of tissue, when fixation is delayed (see, e.g., Bullough, 1950; Graem and Helweg-Larsen, 1979).

Since tissue culture of blastema explants is currently being employed (Stocum, 1968a; Vethamany-Globus and Liversage, 1973a; Globus and Liversage, 1975a,b; Globus and Vethamany-Globus, 1977; Globus, 1978; Vethamany-Globus et al., 1978; Carlone and Foret, 1979; Conn et al., 1979; Globus et al., 1980; Rathbone et al., 1980; Tomlinson et al., 1981) as a means of understanding cause and effect relationships in growth and differentiation of regenerates, it is important to understand and document some of the relevant nuances of culture techniques and their application to regeneration studies. We recently demonstrated (Tomlinson et al., 1982) that subsequent to excision and explanation of the limb blastema into culture medium, there is an almost immediate and precipitous decline in mitotic activity lasting several hours. Coincident with the disappearance of mitosis, abnormal mitotic figures, lacking the condensed chromosomal nature of normal figures, are in evidence and persist until normal levels of mitotic activity are restored approximately 6 hr later. Although cells with unusual nuclear morphology have been observed in sectioned blastemas (pale and vacuolated cells; see Bryant et al., 1971; Tassava and McCullough, 1978; and pseudomitotic figures, see Chalkley, 1954), their significance has not been determined. Abnormal mitotic figures appear as early as 15 min after excision of the regenerate, suggesting that they are likely a product of isolation from the vascular supply and not a result of culture conditions per se. We have suggested that since new supply routes must be established by diffusion of medium into freshly explanted tissues, a transient interruption in the supply of adenosine triphosphate (ATP) to condensed chromosomes could result in uncoiling or decondensation, leading to the formation of abnormal mitotic figures. Lampbrush chromosomes have been shown (Karsenti and Gounon, 1979) to assume a relaxed state and become "fuzzy" or indistinct in the absence of ATP or in the presence of nonhydrolyzable ATP. Moreover, microtubule assembly, chromosome movement along spindle fibers and microtubule disassembly are all ATP-dependent processes (Margolis and Wilson, 1979). If this is the case, then it is conceivable that many of the prophase figures could regress (by decondensation of chromosomes) to the dispersed chromatin pattern typical of interphase cells, so that they are no longer recognizable as prophase mitotic figures (Tomlinson et al., 1982). Presumably, as new supply routes are established by diffusion, chromosomes recondense (see also Brinkley et al., 1980; Karsenti and Gounon, 1979), abnormal mitotic figures decrease in number and mitotic activity approaches preexplantation values. In view of this transient loss of mitotic activity, short-term studies of mitotic activity in cultured regenerates must be interpreted with caution.

Identification of Nerve-Dependent Event(s) in the Cell Cycle

There is general agreement that the primary neurotrophic function in limb regeneration is to facilitate the progression of blastema cells through the cell cycle. The durations of blastema cell cycle phases have been determined in vivo (Grillo, 1971; McCullough and Tassava, 1976; Wallace and Maden, 1976) and in vitro (Tomlinson et al., 1984); however, no consensus has so far been reached with respect to the effects of nerve deprivation on cycling cells. Nevertheless it has been possible to formulate some hypotheses suggesting that neurotrophic factors may influence events in G_1 (Globus, 1978; Maden, 1979a; Olsen and Tassava, 1984) or G_2 (Tassava and Mescher, 1975) phases of the cell cycle, although they are not mutually exclusive (Globus, 1978; Tomlinson et al., 1984).

Resolution of the precise site of action of the neurotrophic factor(s) has been difficult for a number of reasons. Whereas the vast majority of cell cycle studies cited in the literature deal with synchronized cell populations, cycling blastema cells are asynchronously moving through the division cycle with merely 1 or 2 percent of the cells in overt mitosis (M) at any one moment in time. In addition, the more or less general assumption that the blastema is comprised of a homogeneous population of cells with respect to nerve dependence has not been helpful from the standpoint of experimental design. Recent evidence suggests that we are likely dealing with heterogeneous subpopulations of blastema cells which may not respond to denervation in a uniform manner and whose requirements for neurotrophic factors may also vary (Globus et al., 1983; Tomlinson et al., 1984; Brockes, 1984). Moreover, Quastler Sherman plots, recently employed to determine nerve dependency in the cell cycle, have inherent limitations in that they yield cell cycle phase durations (rates of progress) exclusively for the cycling populations of cells. A reduction in mitotic activity consequent to denervation can theoretically be achieved by influencing the rate of cell progression through one or more phases of the mitotic cycle or by selectively reducing the number of cells engaged in cycling (a subpopulation arrest). Quastler Sherman analysis fails to reflect subpopulations of cells that are not continuously cycling and are quiescent (Tomlinson et al., 1984), perhaps in the G_0, G_1, or G_2 phases of the cell cycle.

Apparently, a different experimental design is required in order to identify cell cycle restriction points, perhaps one that is similar to the cell cycle suppression/restimulation strategy employed by Gelfant (1977) and Tomlinson et al. (1984). Organ culture affords the opportunity to resupply nerves to previously denervated regenerates in anticipation of a resumption of blastema cell cycling. Studies employing this kind of strategy will undoubtedly lead to new insights into the neural dependence of the blastema.

Substance P: A Neural Peptide as Candidate for Neurotrophic Factor

The identity of a neurotrophic factor(s) capable of promoting growth of newt limb regenerates has thus far remained elusive. However, the available evidence, largely based on infusion of partially purified nervous tissue extracts into denervated blastemas, indicates that trophic activity resides in peptides of relatively low molecular weight, possibly less than 500 daltons (Jabaily and Singer, 1978; Singer, 1978). Considerable attention is currently being given to polypeptide growth factors such as fibroblast growth factor (molecular weight approx. 13,000), which has been shown to be mitogenic for a variety of mammalian cells and appears to stimulate blastema cell proliferation (Gospodarowicz et al., 1978b; Mescher and Gospodarowicz, 1979). However, there is a strong likelihood that there are more than one and possibly several neurally derived mitogenic factors capable of promoting blastema cell proliferation (see Chap. 6). We recently reported (Globus et al., 1983) that the undecapeptide substance P may be a strong candidate for a neurotrophic role in limb regeneration.

Substance P (SP), first extracted from equine brain (von Euler and Gaddum, 1931) is widely but unequally distributed throughout the central nervous system (Brownstein et al., 1976; Takahashi and Otsuka, 1975), with high concentrations of immunoreactive SP in the substantia nigra and hypothalamic regions of the brain (Powell et al., 1973; Leeman and Hammerschlag, 1967). It is contained in high concentrations within sensory neurons, localized to neuronal perikarya in dorsal root ganglia, to neuronal processes within the dorsal horn of the spinal cord (Hokfelt et al., 1975), and within terminal axons of sensory neurons (Pickel et al., 1977). Interestingly, the concentration of bioassayable SP (or that estimated by radioimmunoassay) in dorsal nerve roots is considerably greater than that in the ventral roots of the spinal cord (Takahashi and Otsuka, 1975), and an analysis of subcellular fractions of rat hypothalamic tissue has shown that it is present in highest concentrations in the synaptosomal fraction (Whittaker et al., 1964). Substance P has been isolated in pure form from bovine hypothalamus (Chang and Leeman, 1970), and the structure elucidated by Chang et al. (1971) was found to have the amino acid sequence H—Arg–Pro–Lys–Pro–Gln–Gln–Phe–Phe–Gly–Leu–Met—NH_2, and an estimated molecular weight of 1340.

Singer's early studies (see review by Singer, 1952) established that, although all nerves are trophic, the sensory supply alone was more effective in promoting limb regeneration than exclusive motor innervation, despite the fact that afferent sensory nerves carry impulses away from the regenerate. This observation led to the suggestion that the trophic function of the nerve was not related to impulse conduction; rather, it was associated with a molecule, probably a protein (Singer et al., 1976), transported in the neuroplasm. Interestingly, electrical stimulation of small-diameter afferent neurons leads to a release of SP (Yaksh et al., 1980).

The general localization of SP in nerve tissues, together with high concentrations of it in sensory ganglia, terminal axons of sensory nerves, dorsal horns of spinal cord, and synaptosomes, is consistent with the localization of neurotrophic activity reported by Singer (1952) and Singer et al. (1976) and strengthens the case for SP as a logical candidate for mediator of neurotrophic effects.

Substance P-Stimulated Mitotic Activity

The ability to test the biological activity of a putative neurotrophic factor is greatly facilitated by the availability of an in vitro assay, primarily because of a decreased likelihood of inactivation of the test substance. In vivo, the molecule may be subject to either rapid turnover or destruction, so that effective concentrations of the test substance may be present for exceedingly short periods of time.

When cone-stage blastemas were cultured on Millipore filters, SP was found to be mitogenic at a concentration of 10 pg/ml of medium (Globus et al., 1983).

Suppression of Mitotic Activity by Antiserum to Substance P

Having demonstrated the mitogenic activity of SP, we looked for further verification, reasoning that if the peptide was immunologically bound and thus inactivated, a suppression of mitosis should occur. Indeed, the mitotic index of ganglionated blastemas was suppressed when they were treated with rabbit anti-SP serum diluted 1:100 in the medium. As the antiserum concentration was reduced to 1:1000 and 1:10,000, a progressive recovery of mitotic activity was evident (Globus et al., 1983); antiserum to myelin basic protein failed to give suppression of mitotic activity. Recent experiments in our laboratory reveal that immunofluorescent SP is predominantly localized in the epidermis of the newt limb blastema with a consistent pattern of staining at the periphery of the epidermal cells, possibly suggesting binding sites for SP. The epidermis remains nonreactive to antibodies prepared against bombesin, met-enkephalin, and neurotensin. Our results to date suggest that substance P may participate in the neurotrophic promotion of blastema cell proliferation. Further elaboration on a substance P role in newt limb regeneration is currently in progress in our laboratory.

Hormone Effects in Vitro

Several hormones have been linked to the regeneration process. Mammalian adrenocorticosteroids, growth hormone (GH), prolactin, and thyroxine have all been shown to promote limb regeneration in hypophysectomized newts (see Schotté, 1961; Wilkerson, 1963; Schmidt, 1968; Connelly et al., 1968; Tassava, 1969; Bromley and Thornton, 1974). In view of the findings that the above-mentioned hormones, as well as adrenocorticosteroid-stimulating hormone, stimulate the production of insulin from the islets of Langerhans (Haist, 1959; Lazarus and Volk, 1962; Tepperman, 1965; Genuth and Lebovitz, 1965; Bern and Nicoll, 1968), the question was raised for the first time as to whether or not insulin is involved in newt appendage regeneration (Fig. 7.2; Vethamany, 1970). In this regard, we also observed that somatostatin, a hypothalamic factor that has a known inhibitory effect on the release of insulin, GH, thyroid-stimulating hormone, prolactin, and glucagon in mammals, also inhibits limb and tail regeneration in the adult newt (Vethamany-Globus et al., 1977). In the past decade, strong evidence has been accumulated, from in vivo and in vitro experiments, that insulin is an essential hormone in newt limb regeneration (Vethamany, 1970; Vethamany-Globus and Liversage, 1973a–c; Vethamany-Globus et al., 1978, 1984).

Our in vivo experiments showed the following: First, the islets of Langerhans in hypophysectomized newts (31 days posthypophysectomy) had undergone extensive atrophy, suggestive of islet inactivity associated with insulin insufficiency. Hypophysectomized animals do not normally feed; starvation together with the absence of ACTH-stimulated glucocorticoids would explain the nonfunctional state of the pancreas. Hypophysectomized newts exhibited an extreme delay or total arrest of limb and tail regeneration (Vethamany-Globus and Liversage, 1973a). Second, sublethal pancreatectomy or alloxan treatment resulting in a diabetic condition in newts also resulted in greatly delayed and abnormal limb and tail regenerates. A few of these diabetic animals exhibiting total arrest in limb and tail regeneration eventually recovered from diabetes and showed signs of reactivation of blastemal formation as late as 44 days postamputation. Subsequent histological examination revealed large hypertrophied islets in the pancreas, strongly suggestive of a correlation between regeneration of the islets and reactivation of blastemal cell accumulation (Vethamany-Globus and Liversage, 1973b). Insulin replacement therapy was not attempted in hypophysectomized newts due to their extreme sensitivity to

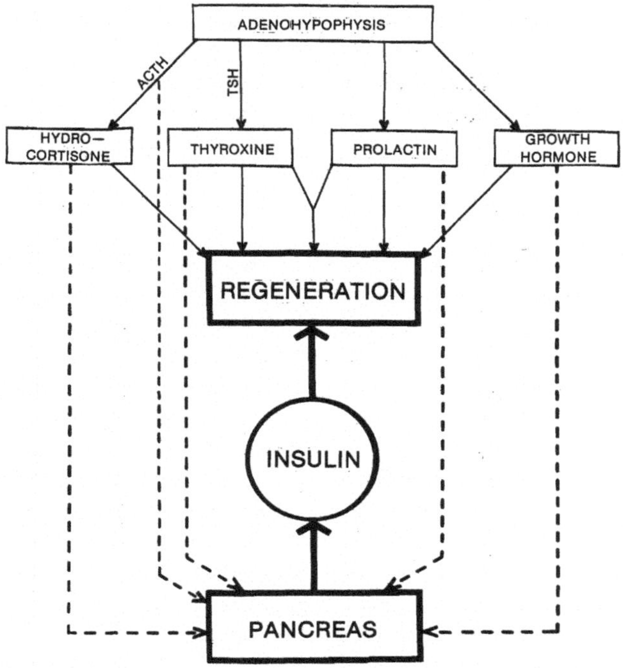

FIG. 7.2. Model showing relationships between anterior pituitary hormones, insulin, and newt limb regeneration. (From Vethamany, 1970.)

the hypoglycemic effects of injected insulin, which would be aggravated by fasting and by the lack of pituitary hormones that mobilize blood glucose (Wurster and Miller, 1960).

In this regard, Sato and Inoue (1973) showed that injections of mammalian GH or an amino acid glucose mixture promoted limb regeneration in hypophysectomized newts. However, it is highly likely that insulin is responsible for the promotion of limb regeneration under these circumstances. Hypophysectomy leads to a depletion of diabetogenic hormones, and the ensuing hypoglycemia reduces the stimulus for insulin production. Replacement therapy with either GH or the amino acid and glucose mixture would reestablish the direct stimulus and restore the production of insulin. Similarly, in newts fasted for three weeks (Tassava, 1983) insulin levels will be low at the time of hypophysectomy, and the latter will result in a severe atrophy of islet tissue (Vethamany, 1970; Vethamany-Globus and Liversage, 1973a) further aggravating the "starvation diabetes" induced by prolonged fasting (see also Haist, 1959). On the other hand, well-fed newts (Tassava, 1983; see also Tassava, 1969) will have higher levels of insulin, blood glucose, and liver glycogen prior to hypophysectomy than fasted newts. There is a lack of trophic stimulation of glucocorticoids in hypophysectomized newts; however epinephrine, glucagon, and insulin-producing systems are still intact and may be induced to operate given the appropriate stimuli. Because of disrupted glucocorticoid-stimulated gluconeogenesis in hypophysectomized animals, a shift to epinephrine and glucagon-induced hepatic glycogenolysis would be expected to occur in order to mobilize blood glucose, and this in turn will trigger insulin production (see Tepperman, 1965). Thus, there exists a strong likelihood that insulin is at least one of the factors responsible for promotion of limb regeneration in well-fed hypophysectomized newts (see also Vethamany-Globus et al., 1984).

In fact, all in vivo experiments involving either the removal of endocrine glands or chemical suppression of hormone production result in widespread derangements in metabolism involving the functions of other organs and glands. Thus, in extirpation experiments it is difficult to determine whether the effects on limb regeneration are primarily due to the lack of the specific hormone or to the deranged metabolism resulting from the hormone deficiency. Results from replacement therapy experiments are also not clearcut, as contaminants of hormone preparations, dosage (the latter leading to hypo- or hyper-hormone condition), and the inevitable interference with homeostasis continue to cloud the issue. The in vitro system permits one to analyze direct effects and the specific responses of the blastema to various hormones.

Effects of Insulin on Cultured Blastemas

Preliminary experiments were done on tail regenerates, simply because they have the regenerating spinal cord as a "built-in nerve source" (Globus, 1970; Vethamany, 1970). The results of a series of our experiments showed that insulin is an essential hormone for growth and cartilage differentiation in cultured tail blastemas in the newt, and, furthermore, a combination of insulin, GH, hydrocortisone, and thyroxine optimizes these effects. Thyroxine also promoted proliferation at low concentrations (10^{-8} M) and cartilage differentiation at high concentrations (10^{-4} M). This provided the first evidence for the existence of a multiple hormone control in regeneration (see Fig. 7.2). On the basis of these results, we proposed that it is not any one hormone that "single-handedly" promotes regeneration of the limb, but it is likely that many hormones are involved (Vethamany, 1970; Vethamany-Globus and Liversage, 1973a–c; Vethamany-Globus et al., 1977).

To study the effects of hormones on the newt limb in vitro, limb blastemas were cultured transfilter to dorsal root ganglia according to the method of Globus and Vethamany-Globus (1977) and the hormones were added singly or in combinations to the culture medium. We have made extensive use of this in vitro system in our attempt to characterize the effects of insulin on the blastema at a macromolecular level.

Blastemal DNA Synthesis and Mitotic Index

Addition of insulin to the culture medium consistently resulted in a significant increase in the incorporation of [^{3}H]thymidine into blastemal DNA in both ganglionated and nonganglionated blastemas. When blastemas were cultured without ganglia, DNA synthesis decreased to approximately 72% of that achieved by ganglionated explants, leaving a substantial level of activity that is apparently nerve independent. When insulin was excluded from the medium, DNA synthesis decreased to 40% of insulin-treated explants, and in the absence of both nerves and insulin it declined to 31% of insulin-treated innervated explants (Vethamany-Globus et al., 1978). A similar insulin stimulation of [^{3}H]thymidine incorporation into blastemal DNA was observed when Lassalle (1983) repeated our in vitro experiments with serum-free culture medium. In addition to confirming our results that insulin significantly increases [^{3}H]thymidine incorporation into blastemal DNA, Mescher and Loh (1981) showed that the insulin-stimulated increase in the incorporation of [^{3}H]thymidine is a function of increasing insulin concentration. Moreover, it is also interesting to note that when blastemas (either ganglionated or nonganglionated) were cultured in the presence of insulin for 48 hr, instead of 12 hr, prior to introduction of [^{3}H]thymidine into the medium, there was a substantial increase in the incorporation of the radiolabel into blastemal DNA (Vethamany-Globus et al., 1978).

Insulin stimulation of [^{3}H]thymidine incorporation into blastemal DNA is paralleled by an increase in blastemal mitosis as well. The overall mitotic index of insulin-treated blastemas is significantly greater than their untreated control counterparts (Vethamany-Globus et al., 1984). We have observed, in our previous work, that the mitotic index of newt limb blastemas, cultured in the absence of nerves, is significantly lower than their innervated counterparts (Globus and Vethamany-Globus, 1977; Globus, 1978). Interestingly, in the absence of insulin the mitotic index of ganglionated explants declined below the basal levels normally observed in denervated, insulin-treated blastemas, and in the absence of both insulin and nerves the mitotic index was depressed further (Vethamany-Globus et al., 1984). The results show that the nerves depend on insulin for the manifestation of the mitogenic effect and emphasize the complementary roles played by insulin and nerves during the phase of rapid proliferation.

Protein Synthesis

In addition to its stimulatory effect on DNA synthesis and mitosis, insulin promoted protein synthesis as well, as evidenced by an increase in ^{14}C-labeled amino acid incorporation into proteins (Vethamany-Globus et al., 1978). Combinations of insulin with GH, L-thyroxine, and hydrocortisone failed to further augment DNA or protein synthesis above the values we observed for insulin alone, which suggests that insulin is not only necessary but, along with nerves, probably sufficient to promote DNA and protein synthesis at the cone stage of regeneration. No appreciable effect on protein or DNA snythesis was induced by exogenous dibutyryl-cAMP, cGMP, theophylline, acetylcholine, sodium azide, or prostaglandin $F_2\alpha$ (Vethamany-Globus et al., 1978).

RNA Synthesis

Recent experiments from our laboratory showed that insulin stimulated the incorporation of [^{3}H]uridine into blastemal RNA in cultures of 6, 12, and 18 hr. Insulin had no effect on RNA turnover, thereby suggesting that the effect of insulin is on synthesis of blastemal RNA (Vethamany-Globus et al., 1984). Withdrawal of insulin from the medium caused a considerably greater inhibition of RNA synthesis than withdrawal of nerves in the presence of insulin.

Effects of Insulin on Incorporation of $^{35}SO_4$ into Blastemal Polysaccharide–Protein Complexes

To date, hormonal influences on blastemal proteoglycans have remained unexplored. In some of the early work we observed that insulin promoted the accumulation of Alcian blue-positive material in the chondrogenic region of tail blastema explants (Vethamany, 1970; Vethamany-Globus and Liversage, 1973c). Results from our recent experiments show that insulin stimulated the incorporation of both [^{3}H]leucine and $^{35}SO_4$ into the total blastemal polysaccharide–protein complexes. The insulin-stimulated increase in $^{35}SO_4$ incorporation was paralleled by a roughly comparable increase in [^{3}H]leucine incorporation, indicating that leucine radioactivity probably reflects incorporation into protein complexed to polysaccharides. Insulin promoted $^{35}SO_4$ incorporation into both chondroitin sulfate and uncharacterized chondroitinase-resistant proteoglycans (Vethamany-Globus et al., 1984).

Blastemal Uptake of Precursor Molecules

We have also observed that insulin significantly increases the blastemal uptake of [^{3}H]uridine, [^{3}H]leucine, $^{35}SO_4$, and nonmetabolizable 2-deoxy-D-glucose, and α-aminoisobutyric acid. In this regard, it has been shown by Singer and Ilan (1977) that denervation does not alter either the rate of transport of amino acids in the postsynaptic cell or the intracellular pool of amino acids available for protein synthesis.

Possible Mode of Insulin Action in Regeneration

From the evidence available from both in vivo and in vitro experiments, it is clear that insulin not only is an essential hormone in newt limb regeneration, augmenting the syntheses of various macromolecules in the regeneration blastema (see Fig. 7.3), but also seems to be essential for the manifestation of the nerve effect (Vethamany-Globus et al., 1984). We had suggested that divalent ions may act as the mediator for insulin action in regeneration blastemas (Vethamany-Globus et al., 1978). We now know that increased

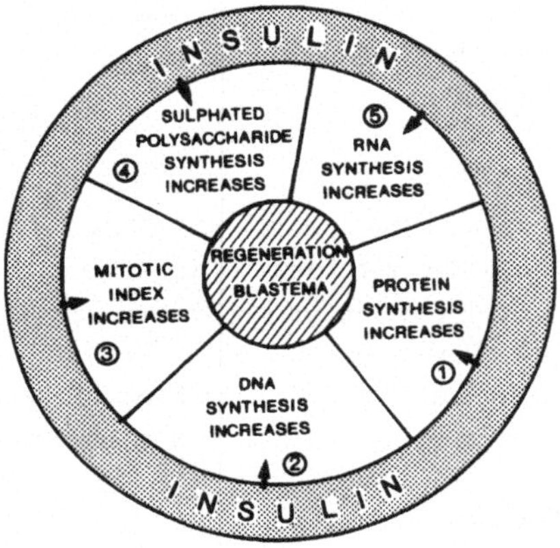

FIG. 7.3. Multiple responses of cultured newt limb blastemas to insulin during proliferative and differentiation phases of regeneration.

intracellular calcium promotes mitosis in newt limb blastemas (Globus et al., 1983). The possibility that intracellular calcium signals may mediate insulin effects on limb regeneration is currently being explored.

How these multiple responses to insulin are coordinated and integrated in the proliferation and differentiation phases remains an enigma. In addition to its effect on DNA replication (S phase), we have speculated that insulin may facilitate the passage of blastema cells through both the G_1 and G_2 phases of the cell cycle (Globus 1978; Vethamany-Globus et al., 1978), by stimulating transmembrane transport of metabolites increasing precursor pools, ultimately stimulating the syntheses of enzymatic and structural proteins as well as nucleic acids required to complete cell divisions. The insulin effect persists during the differentiation stage, as evidenced by the synthesis of blastemal glycosaminoglycans.

From the current state of our knowledge, it is difficult to explain how insulin brings about these multiple responses; however, the scheme Tata (1976) presented for the action of polypeptide hormones is an attractive one. Accordingly, these responses can be grouped into early rapid metabolic responses and late slow responses, the latter connected with the synthesis of specific proteins or overall growth. The rapid effects include events such as enhanced uptake of ions, amino acids, and nucleosides, and alterations of intracellular cAMP, cGMP, and Ca^{2+} levels. Late events include accelerated rate of synthesis of all RNAs, especially ribosomal RNA, stimulation of RNA polymerases, higher concentration of polysomes, enhanced amino acid incorporation, redistribution of free and membrane-bound ribosomes, accelerated assembly of membranes of the endoplasmic reticulum, and other events leading to growth and differentiation (see also Kissebah and Tulloch, 1976).

It is generally held that the multiple actions of hormones arise from an interaction between the hormone and a single species of receptor. For the polypeptide hormones the receptors would occur on the cell surface (Cuatrecasas, 1974). In this regard, Goldfine et al. (1977) have provided evidence that insulin in fact enters intact target cells and then binds to specific nuclear proteins, although insulin does not bind to DNA, RNA, or histone.

The search for specific insulin receptors in blastema cells is currently under way in our laboratory.

Re-Evaluation of Hormone Research in Newt Limb Regeneration

Recently, Hinterberger and Cameron (1983) have also reported that medium supplemented with insulin, thyroxine, growth hormone, hydrocortisone, and chick embryo extract supports extensive blastema cell proliferation and differentiation of both cartilage matrix and myotubes. They emphasized a requirement for embryo extract in addition to the four hormones we had used in our previous work; however, it should be noted that our tail and limb blastema explants were covered by epidermis and innervated by nerves (Vethamany-Globus and Liversage, 1973c; Globus and Vethamany-Globus, 1977; see also Globus, 1978). The blastema fragments Hinterberger and Cameron (1983) used lacked both epidermis and innervation. Thus there is a strong likelihood that one of the functions of embryo extract in their cultures may be to serve as a replacement for factors normally provided by nerves and the epidermis.

In both in vivo and in vitro studies concerning hormones in newt limb regeneration there exists a need for clarification regarding "physiological levels," because the latter can be misleading when strictly applied in vitro. The physiological level of a hormone is a measurement at a particular point in time. In vivo, insulin is secreted in response to blood glucose, in a constant attempt at maintaining homeostasis. Taking into account the fluid dynamics of blood circulation over a period of time, the blastema cells, in vivo, would bind greater amounts of the hormone than the physiological levels might suggest. The physiological level of serum insulin as reported by Liversage and Banks (1980) reflects neither the total amount of insulin available to the blastema nor the amount blastema cells would bind during a period of 2–4 days (our culture period). There exists in vitro a lack of constant replenishing of the hormones to the cells, which is so effectively accomplished in vivo by circulation, and for these reasons physiological levels cannot be strictly adhered to in vitro. Moreover, the concentration of insulin we used in our in vitro experiments, namely, 0.14 IU/ml of culture medium, falls within the dose–respone curve, established for insulin stimulation of mitosis in cultured chick fibroblasts (Von Haam and Cappel, 1940), and more recently [^{3}H]thymidine incorporation into blastemal DNA in cultured newt limb blastemas (Mescher and Loh, 1981; Lassalle, 1983). It seems that the concentration we have used is well below the receptor saturation point.

Furthermore, there seems to be a general expectation that elevations in hormone levels must occur before a hormone can be credited with a regeneration-promoting role. As measured by Liversage and Korneluk (1978) and by Liversage and Banks (1980), the levels of serum thyroxine and insulin do not fluctuate significantly during different stages of newt limb regeneration. However, the hyper-hormone condition need not be a prerequisite for hormone action in limb regeneration. In a homeostatic mechanism with a built-in negative feedback system, there is always a tendency to avoid extreme fluctuations in the levels of hormones such as insulin and thyroxine. The availability of a particular hormone within its physiological range may be sufficient for the blastema cells to bind the hormone to their receptors and for the hormone-mediated specific functions to be initiated and carried out. Furthermore, they refer to the physiological levels of hormones as "background" or "residual" and subscribe to the existence of a "balanced hormonal milieu" during regeneration (Liversage and Brandes, 1982). In view of our current knowledge on hormone action, both in the blastema and in other systems, these terms, unfortunately, undermine the specificity of known cell–hormone interactions and thus are no longer adequate to explain hormone action during newt limb regeneration.

In search of a "particular hormone" that can be credited with a regeneration-promoting role, there appears to be a general tendency to dismiss the blastemal responses to essential hormones, such as insulin, GH, thyroxine and prolactin, and to view the hormone actions as mere manifestations of their abilities to stimulate cellular metabolism (Wallace, 1981). Information regarding cellular responses to various hormones during different stages of regeneration is undoubtedly relevant. For instance, it is likely that hormones such as corticosteroids may be involved in initiation and/or dedifferentiation and that others (insulin, GH, thyroxine and prolactin) may act during differentiation phases. Thus the regenerating limb may be viewed as a system in which the responses to a particular hormone vary according to the specific event(s) occurring during different stages of regeneration; the availability of that particular hormone within physiological levels may be sufficient for limb regeneration to proceed.

Role of the Apical Epidermal Cap

The apical epidermal cap (AEC) also appears to play an important role in control of the cell cycle. During the early stages of limb regeneration, the epidermis migrates distally over the amputation surface and thickens to form the AEC, while beneath the cap mesenchymatous cells accumulate and then rapidly divide to form the regeneration blastema. In order for limb regeneration to occur, it is imperative that direct contact between the AEC and subjacent mesenchymal tissues be established and maintained (see reviews by Faber, 1971; Singer and Salpeter, 1961; Stocum, 1975a; Thornton, 1968). The manner in which the apical ectoderm exerts its effect on the underlying mesenchymal tissues is still unclear; however attention has centered around a cell cycle-promoting role (Globus, 1978; Globus et al., 1980; Tassava and Mescher, 1975). There is a clear indication from the evidence given below that proliferation is favored in the presence of the AEC and cartilage differentiation is manifested in its absence. (For additional discussion of the role of the epidermis in regeneration, see Chap. 3.)

Evidence for Cell Cycle Promotion

The experiments about to be described emerged from preliminary observations (Globus and Vethamany-Globus, 1977) revealing that if the AEC or a portion of it is inadvertently removed from the blastema prior to culture, the mitotic activity of subjacent mesenchymal cells declines and cartilage nodules are precociously formed (Globus et al., 1980). This was reminiscent of our parallel findings in the chick embryonic limb (Globus and Vethamany-Globus, 1976), in which an intact apical ectodermal ridge, cultured in contact with a high-density monolayer of limb mesoderm cells, promoted mesodermal outgrowth, whereas, in the absence of the apical ectodermal ridge, cartilage nodules formed instead. These results have recently been confirmed by Reiter and Solursh (1982).

Effect of Removal of AEC

When blastema explants at the mid-cone stage of regeneration were cultured transfilter to sensory ganglia with intact AEC or with AEC removed by means of treatment with ethylenediaminetetraacetic acid, the decline in both DNA synthesis and mitotic index in the mesenchymal cells of epidermis-free explants was striking, strongly supporting a cell cycle-promoting role for the AEC. What is particularly intriguing is the fact that the decline in DNA synthesis and mitotic activity coincides with the precocious differentiation of cartilage nodules at the apical extreme of the explant (Globus et al., 1980).

Re-Establishment of AEC over Proximal Blastemal Mesenchyme

It seems apparent that the retention of mesenchymal blastema cells in the cycling state requires the presence of the AEC. On the basis of these findings it could be predicted that proximal mesenchyme, at the stage of imminent cartilage differentiation, is no longer in close contact with the AEC and would be expected to exhibit a decline in mitotic activity. In order to test our hypothesis (Globus et al., 1980), a fresh wound epidermis was allowed to be re-established over proximal mesenchyme exhibiting no overt signs of cartilage differentiation. The experimental design is summarized in Figure 7.4. In the experimental group, the distal portion of the blastema was surgically removed at the beginning of the experiment, and the limbs were allowed to continue to regenerate for 2, 3, or 4 days. At the termination of the experiment, the formerly proximal (now distal) part of the blastema with re-established epidermis was excised for analysis and compared with the proximal portion of the

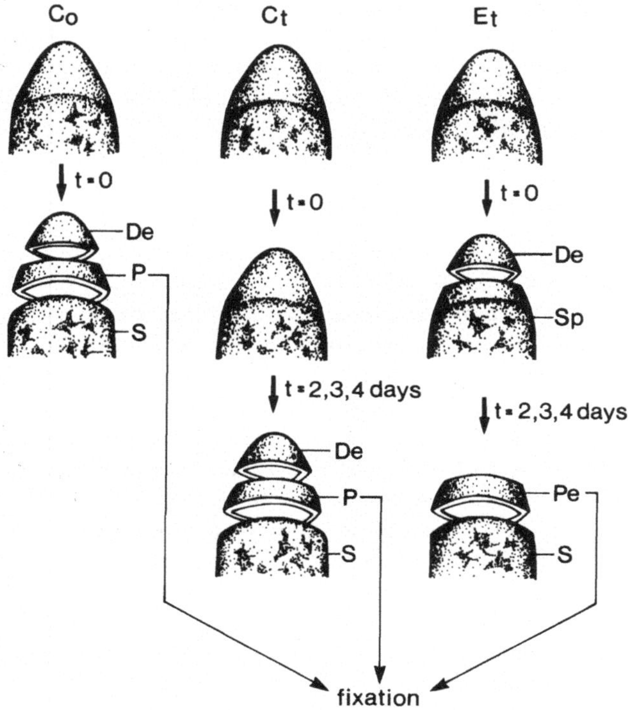

FIG. 7.4. Flowchart summarizing the procedures followed for re-established epidermis experiments in vivo: E_t is the experimental group, where t is the time in days after the excision of the distal portion (De) of the blastema (b); C_t is the control group, where t is time after excision of De from the E_t group; C_0 is the time-zero control, where the proximal portion (P) of a cone-stage blastema (no apical epidermis) is fixed at the start of the experiment; Pe is the proximal part of cone-stage blastema with re-established apical epidermis; S is the forelimb stump; and Sp is the forelimb stump with the proximal portion of blastema intact. (From Globus et al., 1980, reprinted with permission.)

intact controls after a corresponding period of regeneration. In the control group, proximal mesenchyme exhibited a progressive decline in mitotic activity and a corresponding advance of cartilage differentiation with time. In the experimental group, mitotic activity progressively increased and cartilage differentiation was suppressed, presumably because formerly proximal blastemal cells remain under the prolonged influence of the re-established AEC and mimic the response of distal mesenchyme.

We have interpreted these results to indicate (see Fig. 7.5) that the intact apical epidermis provides the "division signal" for cycling cells, so that they may progress, once again, through the mitotic cycle, whereas in the absence of the AEC postmitotic cells cease to cycle and differentiate instead (see also Globus, 1978; Globus et al., 1980; Mescher, 1976; Tassava and Mescher, 1975). Evidence supporting the view that apical epidermis suppresses cartilage differentiation and favors cell proliferation in the mesenchyme subjacent to it can be derived from numerous regeneration experiments (Goss, 1956a,b; Faber, 1960; Michael and Faber, 1961; Thornton, 1968; Mescher, 1976; Tassava and Loyd, 1977). The most attractive prospect of this work relates to the role of the AEC in controlling the developmental program of the blastema cells. When differentiation is imminent, re-establishment of the wound epithelium over proximal mesenchyme results in suppression of cartilage formation and expression of cell division tendencies. Conversely, at a stage (mid-cone) when proliferation is prominent in the blastema, removal of the AEC results in a suppression of cell divisions and precocious promotion of cartilage differentiation. We find it reasonable to view this information in the following way: Since cartilage differentiation first appears in the proximal region of the blastema (and developing embryonic limb) and then progresses proximodistally, it is likely that the AEC influence on the mesenchyme exists in a distoproximal gradient,

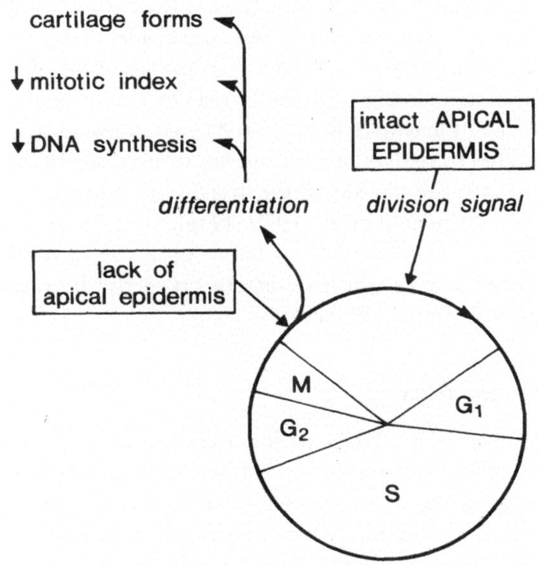

FIG. 7.5. Model illustrating the role of the apical epidermis in the "tripartite control" (Globus, 1978) of cell cycle events in cone-stage limb blastemata in the newt (*Notophthalmus viridescens*. The developmental pathway (i.e., division or redifferentiation) taken by the dedifferentiated blastema cells is influenced by the apical epidermal cap (AEC)." The intact AEC provides the "division signal" for cycling cells, so that they may progress once again through the cell cycle, whereas in the absence of the AEC, postmitotic cells leave the cell cycle and differentiate. (From Globus et al., 1980, reprinted with permission.)

providing the strongest division signal (Globus, 1978) in the region immediately subjacent to the cap, with a proximal diminution of the signal. During the growth of a regenerate, the distal mesenchyme progressively assumes a proximal position so that the influence of the AEC gradually diminishes. The concentration of epidermal factor(s) in the proximal mesenchyme is no longer effective in retaining the cells in the mitotic cycle and they undergo differentiation. More distal cells would receive an effective concentration of the factor(s) and continue to divide until they, too, are displaced to a more proximal position and differentiate. Accordingly, differentiation would progress in the usual proximodistal manner, with proliferation continuing in the subapical mesenchyme. The manner in which the AEC exerts its effect has not been determined.

Role of Calcium

In view of our present state of knowledge, the proliferation of animal cells will undoubtedly involve a multifactorial regulatory chain of events with both extracellular and intracellular mediators, each providing a link between initiation and overt mitosis. Neural, hormonal, and epidermal influences, which we have referred to as the "tripartite control" of proliferation (Globus, 1978), may be considered to be primary signals; however, many cellular responses to external stimuli are at least partly mediated by changes in cation (calcium) flux. Although generated by exogenous mitogens, an increase in intracellular Ca^{2+} is thought to be an intracellular signal for division in many cells (Berridge, 1975; Rasmussen, 1975; Whitfield et al., 1979; Nagle and Egrie, 1981). The importance of Ca^{2+} in limb regeneration has been recognized (Vethamany-Globus et al., 1978); however, it received little attention in the literature until recently (Globus et al., 1983; Sicard, 1983c).

Effects of Calcium Flux

In order to achieve a net change in intracellular Ca^{2+} one might either alter exogenous Ca^{2+} levels or induce a flux in Ca^{2+} across the cell membrane, thereby increasing or depleting intracellular levels. The manipulation of exogenous levels of Ca^{2+} did not yield significant alterations in blastema cell proliferation (Globus et al., 1983). The second option, experimentally induced cation flux, was much more productive.

Calcium Influx Promotes Mitosis

Ionophore A23187 is frequently employed as an experimental tool to transfer Ca^{2+} across biological membranes and cause an increase in intracellular Ca^{2+} levels, that is, a net influx. It has been suggested (Luchasen et al., 1974) that the ionophore behaves as a mitogen in this regard. When cone-stage blastemas were cultured transfilter to sensory ganglia in the presence of A23187, a more than twofold increase in mitotic index was recorded. These results were verified in vivo, where the ionophore was injected intraperitoneally twice daily over a period of 3 days, suggesting that ionophore-induced Ca^{2+} influx is mitogenic (Globus et al., 1983).

An increase in intracellular Ca^{2+} is known to lead to the formation of an active Ca^{2+}–calmodulin complex (see Means and Dedman, 1980), which then regulates protein kinases active during cell divisions. The phenothiazine antipsychotic drug chlorpromazine (Chl) is thought to bind to the intracellular calcium receptor calmodulin (CaM) and cause inhibition of several enzyme systems regulated by CaM (Levin and Weiss, 1979; Weiss and Wallace, 1980). The injection of Chl into newts at an early cone stage of forelimb regeneration suppresses mitosis. Furthermore, when ionophore-promoted Ca^{2+} influx is followed by Chl inhibition of the active Ca^{2+}–CaM complex, the mitogenic effect is abolished. We have recently observed that elevation of cGMP and a suppression of cAMP accompany the ionophore-promoted cell proliferation. On the other hand, Chl causes a suppression of cGMP and an elevation in cAMP coinciding with inhibition of blastema cell mitosis (M. Globus, S. Vethamany-Globus and A. Kesik unpublished). These results extend the earlier findings of Sicard and DiNicola (1974), that Chl injections cause gross retardation of limb regeneration.

Chlorpromazine is also known to function as an antagonist to dopamine and thereby mimics the effect of dopamine depletion. Although we have conducted only very preliminary experiments, our findings (Globus et al., 1983) indicate that dopamine promotes mitotic activity and probably warrants further extensive investigation.

Calcium Efflux Suppresses Mitosis

Our experiments in vivo with papaverine, which promotes a net Ca^{2+} efflux, lowering intracellular Ca^{2+} levels (Duprat and Kan, 1981), resulted in the suppression of blastema cell mitosis and lend support to the idea that changes in intracellular Ca^{2+} may trigger the internal regulation of mitosis during amphibian limb regeneration.

Calmodulin Levels in the Blastema

Upon Ca^{2+} binding, conformational changes in calmodulin occur, allowing it to associate with CaM-regulated protein kinases that lead to the initiation of mitosis. Nagle and Egrie (1981) localized calmodulin in the mitotic apparatus of dividing tissue culture cells, and many others, including Berridge (1975), Cheung (1980), and Chafouleas et al. (1982), have ascribed an important role for CaM in the cell cycle. Recently Chafouleas et al. (1982) have shown that CaM levels are elevated twofold at late G_1 and/or early S and are maintained throughout the remainder of the cell cycle until cytokinesis. The G_1 daughter cells then contain half of the intracellular CaM found prior to cell division. The question then arises as to whether CaM levels change during Ca^{2+}–induced mitotic stimulation. Radioimmunoassays (Globus et al., 1983) of CaM levels in control mid-cone-stage newt blastemas and those treated with either ionophore (stimulatory) or Chl (inhibitory) indicate that CaM levels per se are not significantly altered in response to the treatments employed. It appears more likely that an increase in the number of active Ca^{2+}–CaM complexes, regulating the enzymes, triggers mitotic activity. It remains to be determined whether the effects of exogenous mitogens (e.g., neurotrophic and hormonal) are mediated by internal calcium signals.

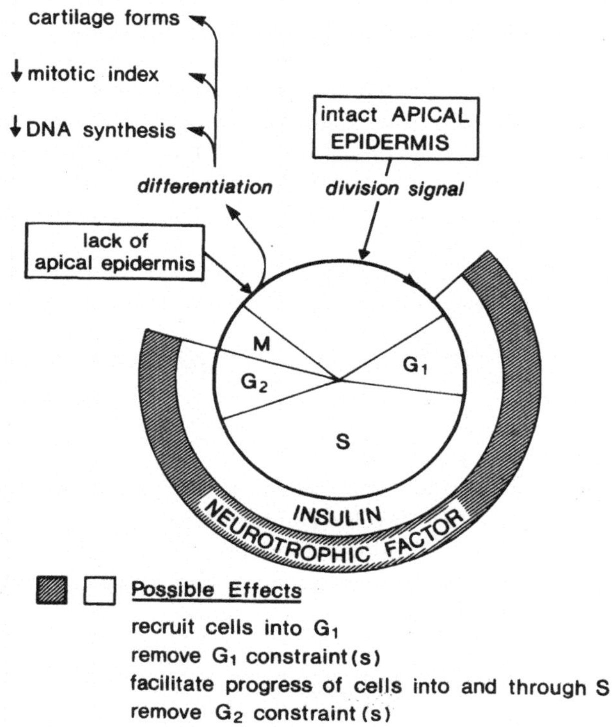

FIG. 7.6. Model illustrating the "tripartite control" of cell cycle events in early to mid-cone-stage limb blastemata in the newt *Triturus*. (1) There appears to be a strong likelihood that nerves have their effect from the G_1 and/or the G_2 phases of the mitotic cycle, although an influence on S has not been ruled out. (2) The developmental pathway (i.e., division or redifferentiation) taken by dedifferentiated blastema cells is influenced by the apical epidermal cap. (3) Insulin appears to play an essential role in DNA replication and may facilitate G_1S as well as G_2M. (From Globus, 1978, reprinted with permission.)

Conclusion

In conclusion, the in vitro methods we have developed have proven to be a convenient and versatile means for investigating the direct responses of regenerating newt limb blastemas to neural, hormonal, and epidermal influences and their interdependence (see Fig. 7.6), without having to contend with the complexities of the whole organism. However, the ultimate necessity to synthesize the information obtained from both in vivo and in vitro experiments must be recognized. Evidence to date from in vivo and in vitro experiments indicates a definite neural, epidermal, and hormonal control in newt limb regeneration. An understanding of the possible interrelationships and interactions among these three major controls and mediators of their actions is just beginning. Many important issues still remain unsolved; however, there is optimism for much progress in this area in the near future.

Acknowledgments

This work was supported by National Science and Engineering Research Council of Canada grant #A6933 to M. Globus, and #A9753 to S. Vethamany-Globus.

8 Leukocytic and Immunological Influence on Regeneration of Amphibian Forelimbs

RAYMOND E. SICARD

During regeneration of the forelimb there is an intimate association between the vascular system and the blastema. This network of blood vessels serves as the channels through which various factors, for example, hormones and nutrients, are conveyed to the blastemal cells. In addition, this network provides the routes over which leukocytes travel to their destinations.

The vascular pattern appears to be distributed in the embryonic limb bud in a fashion that suggests an important role in morphogenesis (Seichert and Rychter, 1971; Caplan and Koutroupas, 1973); however, it cannot yet be determined whether this vascular pattern plays a determinative role in pattern formation during limb development. Nonetheless, the vascular pattern is clearly not fortuitous and, most assuredly, creates a permissive environment for the expression of events of embryogenesis. Whether the vasculature plays a comparable role during limb regeneration is even less clear at present.

A case cannot be made, therefore, for a special role for blood vessels during limb regeneration; however, leukocytes definitely perform an important function. In view of the severity of the wound created by amputation of the forelimb (which serves as the triggering event for regeneration), a rather strong reaction on the part of phagocytes (granulocytes and monocytes) and of immunocytes (lymphocytes and macrophages) can be expected.

Leukocytes as Precursors of the Regeneration Blastema

The abundance of various leukocytes in the wound area early after amputation followed by their disappearance coincident with the emergence of a blastema misled some previous investigators into considering that the blastema had a hematogenic origin (Fritsch, 1911; Hellmich, 1930, 1931; Ide-Rozas, 1937). Favorite candidates for transformation into blastemal cells were lymphocytes and macrophages (MPs). According to Maximow's scheme (discussed by Hellmich, 1930), small and large lymphocytes would transform (or differentiate) into "polyblasts," which, in turn, could further be transformed into "Regenerationzellen," along with numerous other "specialized cells." Among the "specialized cells" that could also contribute to this "Regenerationzelle" population were fibrocytes (fibroblasts), mast cells (basophilic granulocyte derivatives), and melanocytes. An alternative scheme proposed that lymphocytes were precursors of fibroblasts, a notion still debated as recently as the mid to late 1960s (Gesner, 1965). Once converted into fibroblasts, these former lymphocytes could now participate with local fibroblasts in forming the regeneration blastema. Macrophages were similarly believed to be able to give rise to fibroblasts (discussed by Ross, 1970) or to directly transform into blastemal cells (Trampusch and Harrebomée, 1965).

That the regeneration blastema is of local and not of hematogenic origin was definitively demonstrated by several studies that were published between 1935 and 1942 (Brunst and Cheremetieva, 1935; Butler, 1935; Butler and O'Brien, 1942; Thornton, 1942). These studies demonstrated that regeneration occurred from amputation through a limited region of shielded tissue following whole-body irradiation, but did not occur from amputated surfaces lying in an irradiated zone of otherwise nonirradiated animals. In the former case, only limited numbers of leukocytes and hematogenic tissues might be expected to have escaped serious radiation damage. At the least, some impairment of the progress of regeneration should have been noted if these cells formed the blastema. In the latter case, most hematogenic cells and tissues would have been unaffected by irradiation, although local tissue would have been seriously affected. With an intact hematogenic complement, regeneration should have proceeded without incident. Therefore, if the blastema were of hematogenic origin, results opposite to those obtained would have been expected.

Premises for Leukocytic and Immunological Influence on Regeneration

Although they do not serve as a source of blastemal cells, leukocytes can still perform critical functions that will affect the progress of regeneration. Consideration that leukocytes and the immune system might influence one or another aspect of limb regeneration derives from two sources: (1) patterns of wound healing in regenerating and nonregenerating limbs and (2) analogies between the blastema and tumors.

Regeneration begins with a period of wound healing, a process that, of necessity, involves the participation of leukocytes and which may invoke immunological activation (Ward, 1980). Since the qualitative nature of wound healing after amputation determines whether or not regeneration will occur (see Schotté, 1961; Schmidt, 1968; Adair and Taban, 1971), it is reasonable to consider the roles of leukocytes in affecting local cellular and molecular events of the wound healing process.

Ordinarily, following limb amputation in the newt, wound healing progresses without dermal participation, thus allowing intimate association between the wound epidermis and the underlying mesodermal tissues. However, following hypophysectomy or denervation, this pattern is altered and one more reminiscent of mammalian wound healing ensues (for reviews see Schmidt, 1968; Adair and Taban, 1971). Under these circumstances, regeneration does not occur. As compared to events normally occurring in the newt after forelimb amputation, this latter pattern of wound healing displays three significant alterations: (1) limited dedifferentiation, (2) increased fibrillogenesis, and (3) no direct physical contact between the wound epidermis and the mesenchymal tissues. Because of their role in wound healing, leukocytes might be expected to play a significant role in determining the extent to which these three events occur.

There appears to be a curious reciprocal relationship between the tendency an organism has to form malignant tumors and its ability to regenerate (Prehn, 1970; Polezhaev, 1972). For example, Seilern-Aspang and Kratochwil (1965) found that injecting tumors into tails of adult newts produced tumors only in 9% of the cases in which the implants were in a "regeneration field" but occurred in 16% of the cases in which they were outside such a field. The tumors outside the "regeneration field" often led to death, whereas those occurring in the field were found to heal spontaneously. More striking, however, was the production of accessory limbs in 40% of the newts Breedis (1952) treated with carcinogens (e.g., methylcholantrene and benzpyrene). In addition, he obtained only 2 true tumors from 500 cases studied. Furthermore, in reviewing the literature on amphibian tumors, Donaldson and Mason (1975) observed that regulation of induced (or implanted) tumors was much more likely to occur if the tissue affected was one that normally participated in regeneration.

In addition, regeneration seems to be more prevalent among organisms lower on the phylogenetic tree. Such organisms are also more likely to possess a less sophisticated immune system. One might speculate, therefore, on a possible relationship between the status of immunological competence and the nature of interaction with an incipient tumor (or blastema). Furthermore, parallels between the regeneration blastema and tumors (Prehn, 1970; Polezhaev, 1972) have led to the suggestion that an appropriate degree of immunostimulation might promote blastemal growth, as has been observed for certain tumors (Spärck, 1969; Prehn, 1971, 1972; Prehn and Lappé, 1971).

Thus it would appear that the participation of leukocytes in inflammatory responses makes them excellent candidates for influencing events associated with the initiation of regeneration. Additionally, the extent to which the blastema can be considered a sarcoma, at least initially, might dictate whether or not the newt's immune system is modulating early events of limb regeneration as well.

Role of Leukocytes in Wound Healing

Details of the morphological aspects of wound healing can be found elsewhere (e.g., Schmidt, 1968; Adair and Taban, 1971) and will not be repeated here. It should be noted, however, that studies of wound healing disclose three significant aspects of the process: (1) wound closure, (2) demolition, and (3) repair (Ross, 1970). If we ignore hemostasis, wound closure may be considered as an epidermal phenomenon, whereas the stages of demolition and repair involve active participation by leukocytes.

Participating Leukocytes and Their Functions

Phagocytic leukocytes (chiefly granulocytes) are the cells characteristically associated with the demolition phase of wound healing. During the early stages of the acute inflammatory response (the first few days), polymorphonuclear neutrophils (PMNs) are the predominant cells associated with wounds (Ross, 1970; Ward, 1980). As phagocytes, PMNs are expected to remove debris and eliminate bacteria that may inadvertently enter the wound; however, through their powerful hydrolytic enzymes (Cline, 1975; Murphy, 1976), PMNs are believed to play an even more significant role in the degradation of intercellular material and damaged cells (Odland and Ross, 1968). Polymorphonuclear neutrophils are not indispensible, however: Their removal from participation in the inflammatory reaction by treatment with anti-neutrophil serum does not affect the overall progress of wound healing (Simpson and Ross, 1972). In part, this can be attributed to the ability of the wound epidermis to assume part of the phagocytic function of the PMNs (Odland and Ross, 1968; Ross, 1970).

Under normal conditions, the degree and duration of PMN involvement would affect the amount of dissolution of intercellular (extracellular) matrix that occurred, as well as the extent of damage to otherwise intact local tissue. The former activity could be instrumental in freeing cells from their matrices for subsequent dedifferentiation in anticipation of blastema formation. The latter activity (associated with chronic inflammation and the formation of granulation tissue) might be considered hostile to blastema formation, inasmuch as it can provoke increased immunological response and increased fibrogenesis leading to scar formation (Ward, 1980).

As the inflammatory response progresses, other leukocytes enter the wound. Among these are MPs and lymphocytes (Ward, 1980). The MPs, as phagocytes, can assist the PMNs in "cleaning" the wound area. Since they possess a lower concentration of hydrolytic enzymes (Cline, 1975; Ward, 1980), they play a less significant role in demolition. However, in conjunction with lymphocytes, they promote events associated with the repair phase (Dumont, 1965; Ross, 1968; Wahl and Wahl, 1979; Ward, 1980; DeLustro et al., 1983). Both MPs and lymphocytes are believed capable of stimulating fibroblast proliferation and collagen synthesis. Consequently, both of these leukocytic classes promote fibrogenesis and ultimately repair and reorganization in the wound area. Since the early formation of fibrotic scar tissue is apparently incompatible with regeneration (see Schotté, 1961), the time and degree to which MPs and lymphocytes participate in the inflammatory reaction could be instrumental in determining whether or not regeneration occurs.

Thus PMNs, MPs, and lymphocytes appear to play significant roles during the demolition and repair phases of mammalian wound healing. Presuming that the amphibian counterparts (newts possess the same leukocytes; see Sicard, 1983a) function analogously, these three leukocyte classes possess the potential for profoundly affecting (in fact, for determining) whether or not regeneration occurs following forelimb amputation.

Regulation of the Repair Phase of Wound Healing

Central to the regulation of the repair phase of wound healing is the regulation of fibroblast proliferation and the regulation of the composition of the extracellular matrix, that is, hyaluronic acid (HA), chondroitin sulfates (CS), and collagen.

Lymphocytes and MPs exert a positive influence on fibroblast proliferation (Dumont, 1965; Ross, 1968; Wahl and Wahl, 1979; DeLustro et al., 1983). In addition, they increase the degree of fibroplasia (Dumont, 1965), and specifically of collagen synthesis (Wahl and Wahl, 1979; DeLustro et al., 1983). It is unclear, however, whether these cells regulate fibroblast proliferation and fibrogenesis through immunological mechanisms or by some other means, for example, "wound hormones" or "trephones" (Carrel, 1921, 1922; Dumont, 1965; Ross, 1968) or through membrane perturbations of target cells (Golub, 1981).

With regard to extracellular matrix materials, Dunphy and Udupa (1955) defined two phases in rat wound healing: (1) a "productive" or "substrate" phase, occurring during the first 5 days, during which mucopolysaccharides are the predominant substances, and (2) a "collagen" phase, extending from 5 days postwounding, associated with increasing collagen concentrations and increasing tensile strength. Mucopolysaccharide content increased to a maximum at 3 days after wounding and then declined, resuming "base-line" concentrations by the 9th day. Collagen concentrations showed a slow initial rise beginning at approximately the 4th to 6th postwounding day and then reached a plateau at approximately 12 days postwounding. Ross (1968) noted that fibroblast numbers did not reach their maximum until approximately day 6. Production of these substances is affected not only by MPs and lymphocytes, but by a variety of hormones (Asboe-Hansen, 1961; Gesner, 1965) and local nutritional and mechanical factors (Dumont, 1965) as well. For example, thyrotropin and growth hormone increase concentrations of HA and CS, while cortisone inhibits collagen synthesis (Asboe-Hansen, 1961). Although these hormones might have a direct effect on fibroblasts themselves, some of their effect might also be mediated indirectly through effects on leukocytes; for example, cortisone decreases numbers of PMNs, MPs (Asboe-Hansen, 1961), and lymphocytes (Gesner, 1965), and thyroid hormones stimulate lymphopoiesis (Gesner, 1965).

There is, in addition, influence eminating from the wound site itself. For example, tissue damage associated with wounding serves to attract, trap, and activate leukocytes at the wound site (Wilhelm, 1965; Spector and Willoughby, 1965; Ward, 1980). Products of these activated leukocytes can also act to amplify the initial response (Wilhelm, 1965; Spector and Willoughby, 1965; Ward, 1980). In addition, locally produced extracellular matrix may also influence the magnitude of the response: HA has been shown to stimulate PMN function (Verge et al., 1982).

Leukocytes therefore regulate the repair phase of wound healing chiefly through their regulation of the expansion (through proliferation) and the activity of the fibroblast population at the wound site. This role of leukocytes in wound healing, as well as being reinforced by other sources, is itself subject to regulation from a variety of sources, as indicated above.

Relevance to Regeneration

A likely means for leukocytes to influence forelimb regeneration is through their roles in the inflammatory response. However, as compared to the activities occurring during mammalian wound healing, in order for regeneration to occur, the following would be required: (1) The period of demolition must be extended and (2) the onset of fibrogenesis must be delayed or grossly attenuated. The former change would be active, while the latter might be passive; however, both would create permissive conditions allowing regeneration to occur.

(Because of the speculative nature of the following discussion, I will not explore ramifications of many of the points raised; rather, the more relevant, or at least more interesting, literature will be discussed in light of mammalian wound healing strategies or the modifications seemingly necessary to allow regeneration to proceed.)

Should the activity of PMNs be allowed to proceed for a longer period, or to a greater extent, the amount of extracellular matrix removed during the period of demolition would be increased. Consequently, a greater number of cells would be freed from their matrices and would become available for dedifferentiation. Delaying the onset of fibrogensis would allow this pool of freed cells to interact more intimately with the wound epidermis; such an interaction might provide the necessary "signal" to bring about this dedifferentiation (see Chap. 3 for a discussion of the role of the wound epidermis). Failure to allow either or both changes to occur would result in a mammalian-type wound healing pattern being retained (or occurring in the case of the newt). On the other hand, causing either or both to occur in a nonregenerator should produce some attempt, albeit abortive, at regeneration.

Hay and Fischman (1961) disclosed that leukocytes migrate into the wound site during the first 5 days after amputation in the newt. Although their numbers do not increase, they remain until day 15. There are no data on PMN function during early stages of regeneration in amphibians; however, high levels of hydrolase (Schmidt, 1968) and collagenase activity (Grillo et al., 1968; Dresden and Gross, 1970) and the synthesis of HA and CS (Toole and Gross, 1971) have been found associated with the stump tissues for significant periods after amputation. In particular, HA synthesis reached its peak about 10 days after amputation and was still significantly above base-line levels 3–4 weeks after amputation. Synthesis of CS did not reach its peak until much later (25–35 days; see Toole and Gross, 1971). In mammalian wound healing, synthesis of these same substances reached their peak by 3 days and had returned to base line by 9 days postwounding (Dunphy and Udupa, 1955). In addition, HA has been shown to *stimulate* PMN function in at least one system (Verge et al., 1982). These data, therefore, would support the possibility of maintaining prolonged active PMN function in matrix degradation. The time frame would be more than adequate to allow for the release of an abundant precursor pool for blastema formation, which is usually apparent by approximately 2 weeks postamputation (see Wallace, 1981, p. 8).

In discussing the restoration of regenerative capacities, Polezhaev (1972) suggested that increased tissue destruction is required. In fact, certain experiments in which hypertonic salt solutions were used to induce "regeneration" in postmetamorphic frogs (e.g., Rose, 1944, 1945; Singer, 1954) might have achieved their success because of such an occurrence. These trials might not have produced more complete regeneration because of a failure to adequately modify events of the repair phase of wound healing as well.

Chalkley's (1954, 1959) detailed analysis of the distribution of mitoses in the regenerating limb disclosed that most of the mitoses were associated with connective tissues (CT) rather than blastemal cells at 7 days (64% among CT and 32% among blastemal cells) and at 13 days (60% in CT and 39% in blastemal cells). In addition, within the CT, only 14% of all mitoses were associated with loose and subcutaneous CT at 7 days, while 36% of mitoses occurred among these tissues at 13 days. At 19 days, CT possessed only 24% of mitoses (15% among loose and subcutaneous CT) and at 25 days only 8% (3% in loose and subcutaneous CT). This increasing proliferation in CT to 13 days postamputation could be explained by increasing activation by MPs and lymphocytes present during that time. Why these leukocytes do not simultaneously induce comparable accumulations of collagen might best be answered by recalling the high levels of collagenase activity found in the regenerating limb during this time (Grillo et al., 1968; Dresden and Gross, 1970).

Since various hormones (e.g., cortisone, thyrotropin, growth hormone) also influence the amount of fibrogenesis (Asboe-Hansen, 1961; Gesner, 1965), additional support for the notion that the degree of firbogenesis determines whether or not regeneration occurs can be derived from studies employing hypophysectomy to suppress regeneration or hormone supplementation to induce regeneration. As shown by Hall and Schotté (1951), a cicatricial scar forms on the stump of hypophysectomized newts following amputation. This operation would deprive the newt of both thyrotropin and adrenocorticotropic hormone (ACTH) (which would stimulate cortisone release) and, in this manner, allow fibroplasia to occur. Hormone supplementation, for example, by administration of ACTH or cortisone to mice (Schotté and Smith, 1961) or by adrenal implantation in frogs (Schotté and Wilber, 1958) might lead to some suggestion of a regenerative response by decreasing or delaying fibrogenesis. More complete regeneration might not occur if an inadequate pool of "blastemal cells" could not be generated.

Yet another means whereby leukocytes could participate in promoting regeneration has been suggested on occasion. This method parallels that observed in the stimulation of fibroblast proliferation by MPs and lymphocytes. In fact, its initial demonstration was that of the stimulation of fibroblast proliferation (Carrel, 1921, 1922). Extending from this early work, it was proposed that leukocytes produced "trephones," nutrients or growth-promoting substances. Such a growth-promoting influence of leukocytes on blastemal cells has actually been proposed (Liebman, 1946, 1949). In addition, these "trephones" were believed capable of transforming affected cells (discussed in Dumont, 1965) and thus might be responsible for dedifferentiation. However, other than the effects on fibroblasts, no convincing evidence has emerged for a unique, general trephocytic function for leukocytes (Dumont, 1965).

The Immune System as Potentiator of Events of Regeneration

The regeneration blastema has been likened to a tumor (e.g., Prehn, 1970; Polezhaev, 1972). This has prompted investigators to use each as a means of learning about the other. Thus it has sometimes been suggested that our understanding of how factors regulate tumorigenesis can be used to gain insight into the events leading to blastema formation, and vice versa. Among those factors considered is the immune system, which has been shown to be able to promote (as well as impair) the growth of tumors under appropriate circumstances (Spärck, 1969; Prehn, 1972).

Analogies Between the Blastema and Tumors

The parallels between the formation of a tumor and the regeneration blastema seemed so great that Rose (1970) actually labeled tumor formation "a case of never completed regeneration." A careful comparison of tumors and the regeneration blastema (and the ways in which they arise) led Prehn (1970) to identify the following characteristics:

1. The blastema and a tumor both are induced in response to trauma and seem to be associated with wound repair. Amputation generally serves to

initiate regeneration; tumors are believed to arise following repeated insult.

2. Blastemas and neoplasias display a relative lack of morphological and histological differentiation. The blastema is considered the product of dedifferentiation while the neoplasia arises from metaplasia.
3. Blastemas and tumors are both characterized by increased growth rates. Neither are believed to possess cells with shortened generation times but, rather, to possess a greatly augmented population of proliferating cells relative to surrounding tissues.

Other similarities that are less universal were also cited:

4. Redifferentiation occurs in both cases, although it is less consistent with tumors.
5. A dependence upon initial nerve sprouting may exist for both blastemas and tumors. The neural dependence of the regeneration blastema is well known (see Chap. 5 for a discussion of this issue). In some cases, nerve sprouting into a future tumor site is noted prior to the appearance of the tumor. Such a situation occurs in rodents following the application of hydrocarbons.

Up to five generalized characteristics appear to be shared by the regeneration blastema and tumors; however, even among these apparent similarities, differences seem to occur. Rose (1970) and Polezhaev (1972) underscore perhaps the most significant difference between them: the difference in how they respond to regulatory factors. Care, therefore, must be exercised when relating information obtained from one system to the other.

Immunostimulation as Promotor of Tumor Growth and Regeneration

In considering the role of the immune system, one invariably thinks of the maintenance of organismal integrity, that is, the discrimination of "self" from "nonself." Tumor cells are considered foreign; therefore we expect them to be eliminated. What does it mean, therefore, when tumor cells are not eliminated and a palpable tumor arises?

One explanation for this occurrence is that one or another prerequisite for effective immunosurveillance (Prehn, 1970) was not satisfied. Among these prerequisites are (1) the early *expression* by the neoplastic cell of a stable surface antigen capable of stimulating the immune system, (2) an immunocompetent host, and (3) susceptibility of the tumor cells to immunological attack.

A second means through which an adequate tumor cell population could be produced is by actual promotion of tumor growth by the immune system itself. Such a mechanism has already been observed (Spärck, 1969; Prehn, 1972); however, its operation depends upon a precarious balance (Prehn, 1970, 1972;

Prehn and Lappé, 1971), and the mechanism through which it operates is not fully understood (Prehn, 1980). This second mechanism has given rise to an immunostimulation theory of tumor development (Prehn and Lappé, 1971). Although its initial documentation is based on immunological influence on the growth of sarcomas (Spärck, 1969; Prehn, 1971, 1972; Prehn and Lappé, 1971), this mechanism has been proposed to perhaps influence regeneration as well (Prehn, 1970). According to this theory, a mild degree of immunostimulation will promote the growth of a responsive tumor (or blastema). If the immunological reaction is too great, rejection or killing of the tumor results; if, on the other hand, the response is too weak, then the rate of tumor growth is not stimulated. However, the proper degree of immunostimulation that promotes tumor growth has not been defined.

These two mechanisms for allowing or promoting tumor growth need not be mutually exclusive. In addition to these mechanisms for stimulating tumor growth, another physical condition must be satisfied before significant growth can occur. This condition is the presence of an adequate vascular supply. Any tumor that reaches palpable mass must be capable of inducing angiogenesis (Prehn, 1980). Without this, the tumor mass remains limited. The presumably well-vascularized nature of the blastema throughout regeneration would suggest that it possesses an abundant capacity to induce angiogenesis.

If the regeneration blastema is truly analogous to a tumor, then its behavior vis-à-vis the newt's immune system should parallel that of tumors. More specifically, four questions must be addressed in determining the nature of immunological influence on regeneration in amphibians: (1) Are amphibians (and in particular urodeles) immunocompetent with regard to tumor rejection? (2) To what extent are the dedifferentiated mesenchymal cells of the incipient blastema antigenic? (3) Are these cells subject to immunological attack? (4) Does the degree of immunostimulation influence the formation and growth of the blastema?

Evidence of Leukocytic and Immunological Influence in Regeneration

The preceding discussion has been primarily theoretical and has endeavored to persuade the reader that there is reason to expect leukocytes and the immune system to play roles in regulating, or at least influencing, the progress of regeneration. This section will discuss what has been discovered attempting to establish and define these roles.

Changes in Leukocyte Levels During Regeneration

Following forelimb amputation, an inflammatory response occurs that involves activation of specific leukocytes (PMNs, MPs, and lymphocytes) and which might stimulate leukopoiesis as well. Thus insight into the role of leukocytes in regeneration might be obtained by assessing quantitative as well as qualitative changes in leukocytes during regeneration.

Sicard (1983a) observed a 50% reduction in circulating leukocytes during the first 2 weeks after amputation. This reduction affected all leukocytes classes, but seemed less pronounced among PMNs. Lymphocytes displayed maximal reduction by 5 days postamputation and did not show recovery until after the 15th day. Macrophages displayed continued reduction until the 15th day post-amputation. Eosinophilic and basophilic leukocytes remained at depressed levels until after the onset of differentiation (20 or more days after amputation). In contrast, PMNs displayed an increase at 5 days postamputation, but were essentially at nonregenerating levels throughout much of the period of study.

These changes in circulating blood cell levels parallel those expected in mammalian inflammatory responses (Erslev and Gabuzda, 1979). Reductions in cell number could reflect preferential migration to the wound area. These changes are also consistent with the observations of Hay and Fischman (1961) showing leukocyte migration into the wound area unil 5 days postamputation with no further increases in numbers thereafter. In addition, they are consistent with Schmidt's (1968) description of the histological picture of the wound site, in particular the indication of unusual numbers of eosinophilic granulocytes at the wound. This latter characteristic is believed by some to denote prolongation of the inflammatory reaction (Hirsch, 1965). Thus the pattern of changes in the circulating leukocyte counts suggests that the functional characteristics of newt leukocytes are comparable to those of mammals. In addition, they suggest the possbility of prolonged local inflammatory activity. No suggestion of any other special or unique role for leukocytes emerges from these data.

In contrast to mammals, newts possess a special class of nucleated cell, the thrombocyte. [At times, these cells can bear a striking resemblance to lymphocytes.] Thrombocyte counts displayed up to 67% reduction during regeneration (Sicard, 1983a). Since there is essentially nothing known about these cells in the newt, nothing definitive can be said of the significance of this observation. One might speculate, however, that the thrombocytes are emigrating from the circulation to the wound site where they might be aiding to stimulate blastemal growth. Such a role is not inconceivable in view of the effects of platelet-derived growth factor on mesenchymally-derived cells (Scher, et al., 1979, Stiles, 1983).

Administration of bovine serum albumin, anti-lymphocyte serum, or cobra venom factor delayed the progress of regeneration (Sicard, 1981a); however, as a group, these failed to produce consistent effects on changes in either leukocyte or thrombocyte levels in the circulation (Sicard, 1983b). In addition, cyclophosphamide suppressed regeneration coincident with a severe leukopenia (up to 80% reduction in leukocytes) (Schotté and Sicard, 1982). However, following cessation of treatment, recovery of regeneration was observed, although the leukopenia was not corrected (Schotté and Sicard, 1982).

These latter data do not support the notion that leukocytes perform an important role in influencing the progress of forelimb regeneration; however, neither can they disprove any such role.

Studies of Lymphoid Organs

The major hemopoietic organs of the newt are the liver, spleen, and thymus (Hightower and St. Pierre, 1971). Granulopoiesis occurs principally in the liver, while lymphopoiesis occurs in the spleen. Unlike mammals, the adult newt (*Notophthalmus viridescens*) does not appear to have a hemopoietically competent bone marrow. Since no reports exist of leukopoietic activity of these organs during regeneration, no additional support can be provided for the supposition, based on circulating leukocyte counts (Sicard, 1983a,b), that leukopoietic responses of the newt to amputation are qualitatively similar to those of mammals.

Besides serving as a source of leukocytes, the lymphoid organs (spleen and thymus) play significant roles in affecting immunological function (McConnell, 1975). Thus removal of the spleen, thymus, or both would diminish lymphopoietic capacity and should impair immunological responsiveness. Fini and Sicard (1980) examined the effects of splenectomy on the progress of regeneration. Splenectomy was chosen rather than thymectomy because of the greater ease and less traumatic nature of the operation and because of the demonstrated effect of splenectomy in suppressing tumor growth (Nouza et al., 1973). Splenectomy followed by forelimb amputation produced a greater reduction in the thrombocyte count, did not affect the number of lymphocytes, and produced an elevation in the numbers of PMNs as compared to normally regenerating newts. Except for a possible acceleration of the time course of regeneration when splenectomies were performed 10–20 days before amputation (in fall or winter, but not spring or summer), no effect on regeneration was observed (Fini and Sicard, 1980).

Recently, we have observed differences in mitogenic responsiveness between splenocytes from wounded newts and from newts undergoing forelimb regeneration (R. E. Sicard, M. F. Lombard, M. D. Albert, unpublished data). Increased responsiveness to both concanavalin A (Con A), a putative T-cell mitogen, and bacterial lipopolysaccharide (LPS), a putative B-cell mitogen, was observed at 2 and 8 days after wounding (removal of the humerus). In contrast, responsiveness to Con A was increased at 2 and 16 days after amputation, while responsiveness to LPS was increased only at 2 days after amputation. Thus, splenocytes from both wounded and regenerating newts displayed strong mitogenic responsiveness to Con A and LPS 2 days after trauma, when acute inflammatory responses are active. Furthermore, splenocytes from newts undergoing limb regeneration showed no further sensitivity to a putative B-cell mitogen and displayed increased responsiveness to a putative T-cell mitogen that was delayed from 8 days after trauma (amputation), when dedifferentiation should be actively ongoing, until 16 days after amputation, when the blastema is well established. Although these data are of a preliminary nature, they nevertheless suggest that important differences in immunocyte (splenocyte) response and behavior exist following wounding that will lead to a regeneration response and wounding that will not. These differences could affect the formation, persistence, and expansion or growth of a presumptive pool of blastema cells.

Attempts at Manipulating Immunological Function

The adult newt possesses lymphocytes that seem functionally to resemble mammalian B- and T-lymphocytes (Ruben and Edwards, 1975). The B-lymphocytes, although competent to produce antibodies, seem to be limited to the production of IgM (Ruben, 1983). The T-lymphocytes appear to be a heterogeneous cell type (Ruben and Edwards, 1975) which includes a "helper" subclass (Ruben and Selker, 1975). In addition, the newt possesses an affective cell-mediated immune response that rejects allografts, although it does so more slowly than in mammals (Cohen, 1966). Thus the newt should be capable of recognizing and reacting to a variety of antigens and should possess the capacity to react against tumors.

Between 1965 and 1970, Taban and co-workers presented a series of reports on immunological control of regeneration (Taban and Charollais, 1965; Taban et al., 1966, 1967a,b; Taban, 1970). These studies showed that rabbit antiserum to a variety of newt tissues, including homogenates of regenerates, adversely affected the rate of regeneration. The active component of the rabbit serum appeared to be the immunoglobulin fraction (Taban, 1970). It is interesting to note that normal rabbit serum also delayed regeneration (Taban et al., 1966) and that rabbit antiserum to homogenates of regenerates often produced less severe retardation of regeneration (Taban et al., 1967b). Although these studies preceded the formulation of the immunostimulation theory, the experimental results are consistent with its predictions. If one assumes that it was the rabbit immunoglobulins that constituted the primary foreign antigens of the rabbit sera, immunostimulation of an adequate magnitude occurred following all treatments, resulting in an adverse effect to regeneration rate.

More recently, Andrews (1976) attempted to test the immunostimulation theory by another means. Adult newts were presented with first or second set allografts, xenografts, a nonspecific immunostimulant (which was not identified), and the vehicle control. Only the allografts and the vehicle control groups showed "greater regeneration." The effects produced by the xenograft and the immunostimulant were not presented. In addition what "greater regeneration" means was never defined. Nonetheless, Andrews concluded that mild immunostimulation was stimulatory and that immunostimulation that was too great was growth inhibitory. The conclusion is consistent with the immunostimulation theory; however, the nature of the information presented only allows the first part of his conclusion to be defended.

Certain of the studies of circulating leukocyte counts cited above (Schotté and Sicard, 1982; Sicard, 1983a,b) also attempted to provide challenges that might alter immunological function and thus test the immunostimulation theory's application to regeneration. Although cyclophosphamide, a potent immunosuppressant, completely suppressed blastema formation, it cannot be concluded that this was the effect of immunosuppression rather than a consequence of another one of the drug's actions (see discussion in Schotté and Sicard, 1982). When cyclophosphamide was used, no blastema formed whether treatment was begun 20 days before amputation or was delayed until 5 days after amputation (Schotté and Sicard, 1982). In the studies using other immunoactive agents, the effects on bud formation were less striking. For example, bovine serum albumin and anti-lymphocyte serum delayed bud formation, as well as later stages. Cobra venom factor, however, only delayed attaining later stages (Sicard, 1981a; Sicard and Laffond, 1983). These results show a consistent effect on blastemal growth, but a variable effect on blastema formation, itself also a manifestation of growth of the blastemal cell population. In striking contrast to these results, a mouse fibroblast-derived interferon *accelerated* bud formation but delayed reaching the terminal stages of regeneration (Sicard, 1981a; Sicard and Laffond, 1983). No explanation is yet available for this particular effect.

Several years ago, R. E. Sicard (unpublished observations) noted an apparent acceleration in the rate of growth (elongation) of blastemas of newts receiving injections of a crude thymus extract powder. Within the past few years a group at Boston College, headed by Allyn Rule, has isolated specific fractions from calf thymus extracts whose biological activities include the stimulation of expression of T-"helper" (TE-H) or T-"suppressor" (TE-S) traits by mouse lymphocytes in culture. These fractions have been used in an attempt to test Prehn's immunostimulation hypothesis with regard to regeneration. At high concentration, TE-S delayed the progress of regeneration; TE-H did not affect bud formation, but delayed reaching subsequent stages of regeneration (Sicard, 1981a; Sicard and Laffond, 1983). These results are also consistent with the immunostimulation theory's predictions.

Although the delay of regeneration obtained with these treatments was modest, TE-S, TE-H, and interferon decreased the numbers of animals that completed regeneration and significantly increased the frequency of morphologically abnormal regenerates produced (Sicard, 1981b; Sicard and Laffond, 1983). The frequency of abnormal regenerates induced by both thymic extract fractions appeared to be dose dependent. Closer examination of the abnormal regenerates disclosed that interferon produced abnormal regenerates whose volumes (mass) did not differ from those of normal controls. The two thymic extract fractions produced opposing results: TE-S increased the mass of the regenerates, while TE-H decreased the mass of the abnormal regenerates (Sicard and Laffond, 1983). Taban et al. (1966) had also observed an increase in the frequency of abnormal regeneration following treatment with rabbit serum that was attributed to diminished blastemal mass. It was especially curious to note that TE-H and TE-S produced principally deletions in the autopodium (Sicard and Laffond, 1983). This occurred whether treatment was begun 18 days before amputation or did not begin until the time of amputation. Preliminary histological analysis suggested a broad spectrum of possible defects. These are currently under investigation.

Our interpretation of the immunostimulation theory of tumor growth led us to expect modification either in the frequency of timing of bud formation or of the regeneration rate; however, changes in blastemal mass are not entirely unexpected results. In this respect, apparently enhanced immune responsiveness (TE-H treatment) clearly did not favor blastemal growth. On the other hand, apparently compromising immune responsiveness (TE-S treatment) seemed to promote blastemal growth, but impaired normal morphogenesis. Such results are not inconsistent with the postulates of the immunostimulation theory. No mechanism can as yet be suggested for producing these effects on regeneration rate, blastemal mass, or autopodial deficiencies (Sicard and Laffond, 1983).

Unfortunately, in none of the studies reported has any aspect of immunological activity in the experimental animals been directly assessed. Therefore there is no evidence that any manipulation of immunological function has actually occurred. Although it might be expected that most of the treatments employed would have provoked some degree of immune response, evidence that any of these proteins were recognized as "foreign" and that an immunological response resulted is lacking. Consequently, although most of the data are compatible with the immunostimulation theory's predictions, the results can neither clearly prove nor disprove the validity of the theory or its applicability to regeneration.

Current Status of the Problem and Directions for Future Study

Leukocytic Influence

It is generally acknowledged that the nature of the wound healing response strongly affects, and may actually determine, whether or not regeneration occurs (Hall and Schotté, 1951; Needham, 1952; Schotté, 1961; Schmidt, 1968; Adair and Taban, 1971). In view of this conviction, it is surprising that greater attention has not been focused on this period of regeneration.

Numerous studies have described alterations in the wound healing patterns of regenerators and nonregenerators, for example, wound healing accompanied by cicatrization following hypophysectomy in adult newts (Hall and Schotté, 1951) and modification of wound healing in frogs receiving adrenal implants (Schotté and Wilber, 1958) or in mice receiving ACTH or cortisone (Schotté and Smith, 1961). All such studies describe conditions that lead to altered wound healing (and the suppression of regeneration or apparent stimulation of regeneration, as the case may be) but do not address the issue of how the wound healing response has been modified physiologically. In particular, the potentially important role of leukocytes has been largely ignored. From histological studies (see Schmidt, 1968, for a brief review) and from data on circulating leukocyte counts during regeneration (Sicard, 1983a), it would appear that leukocytic responses to wound trauma in newts is essentially comparable to what is seen in mammals (Ward, 1980); however, there are no physiological data to support this premise. Consequently, we do not know if wound healing conditions that favor (or prevent) regeneration are the result of activities of local tissues (e.g., connective tissues) or are brought about by the leukocytes participating in the wound healing process. Certain preliminary observations (R. E. Sicard, M. F. Lombard, M. D. Albert, unpublished data) suggest that differences in splenocyte behaviors following the acute phase of inflammation might contribute, at least in part, to enabling a pool of presumptive blastema cells to emerge following amputation.

Schmidt's observation that "systematic study of inflammation in the amputated limb of the adult newt should prove well worth while" (Schmidt, 1968, p. 24) bears reaffirmation; however, future studies must include evaluation of leukocyte functions. Characterization of the physiological responses of PMNs, MPs, and lymphocytes during the early stages of regeneration (the first week after amputation) and under conditions that suppress regeneration should prove especially enlightening.

Immunological Influence

The suggestion of immunological influence on regeneration is a recent one (Prehn, 1970). This notion, deriving from apparent similarities between the regeneration blastema and tumors, suggests that mild immunostimulation will enhance blastemal growth but that too much immunostimulation will impair it. Following appropriate immunological manipulation, we would therefore expect to observe a change in one or more of the following: (1) the time required to form the blastema, (2) the rate of regeneration, and (3) blastemal mass.

A variety of studies (e.g., Cohen, 1966; Ruben and Edwards, 1975; Ruben and Selker, 1975; Ruben, 1983) indicate that the newt's immune system, although perhaps not as complex, appears to parallel that of mammals; however, nothing is known of its functioning during regeneration.

As indicated earlier, there are four questions that need to be addressed in assessing immunological influence on regeneration. From Seilern-Aspang and Kratochwil's (1965) demonstration of the "spontaneous healing" of tumors, it would appear that the newt is immunocompetent with regard to tumor rejection. Recently Kintner and Brockes (1984) demonstrated that monoclonal antibodies can be produced by mouse lymphocytes against blastemal cells that are different from those produced against skeletal muscle. Furthermore, the antiblastemal antibodies are IgM [the only class that newts are known to produce (Ruben, 1983)], whereas the anti-muscle antibodies are IgG. However, we still do not know the extent to which the dedifferentiated mesenchymal cells of the blastema are antigenic to the newt or whether these cells are subject to immunological attack by the newt's own immune system. Concerning the influence of the degree of immunoreactivity on the formation and growth of the blastema, certain studies (Andrews, 1976; Sicard, 1981a,b; Schotté and Sicard, 1982; Sicard and Laffond, 1983) suggest that Prehn's (1970) immunostimulation theory may apply to regeneration. In addition to altering the time and frequency of blastema formation, certain treatments used in these studies led to morphogenetic defects (Sicard and Laffond, 1983).

A variety of immunoactive agents have been found to affect regeneration in ways that are consistent with the immunostimulation theory of Prehn (1970). More detailed systematic analysis of histological and morphological changes associated with these treatments is warranted. In addition, drawing on our expanding understanding of amphibian immunity, assessment of immunological activity in the newt during regeneration must be undertaken. Evaluation of parameters relating to cell-mediated immune response and tumor rejection would be particularly important in this regard. In addition, specific modification of these functions by substances previously used (especially cobra venom factor, TE-H, TE-S, and interferon) must be demonstrated. Furthermore, the dedifferentiated blastemal cells should be closely examined to determine their ability to (1) provoke an immunological response in newts and (2) be subject to immunological attack by the newt's own immune system. Results of such studies should prove enlightening in appreciating the role of the immune system (and of blood cells), not only in regeneration but also in development, more generally.

Summary

Participation by leukocytes in regeneration has been suggested from time to time. Leukocytes have been excluded as the source of blastema cells; however, their activities, especially during wound healing, might be instrumental in determining whether or not regeneration occurs. This could be accomplished by modulating activities of local connective tissues during the demolition and repair phases of wound healing, thereby affecting the degree of fibrogenesis and fibroplasia, by promoting the growth and proliferation of blastema cells, or by affecting both of these events. More recently, analogies between the blastema and tumors have led to the suggestion that the immune system might influence regeneration. This notion proposes that mild immunostimulation is stimulatory, whereas weaker or stronger immune stimulation has adverse effects on the progress of regeneration.

Interest in this area has not been great; therefore, the possible roles of leukocytes and of the immune system remain unconfirmed. Evidence that has accumulated suggests that (1) functional differences in immunocyte behaviors exist between newts undergoing regeneration and newts healing a wound only, and (2) putative immunological manipulations adversely affect the progress and final outcome of regeneration. These observations are consistent with the general premises, although not necessarily with the particulars, of the prevalent hypotheses for leukocytic and immunological influence on regeneration. Moreover, these results suggest that there might be influences emanating from these sources affecting both the initiation of regeneration and morphogenesis.

The large number of unaddressed and unanswered questions in this area prevents us from properly understanding the roles of leukocytes and of the immune system in regeneration and development. On the other hand, the intriguing results that are beginning to emerge suggest that this should be a fruitful area for future studies.

9 Synopsis and Conclusion

RAYMOND E. SICARD

The preceding chapters have surveyed several aspects of the regulation of vertebrate limb regeneration. In particular, they have (1) identified major events of limb regeneration that are influenced by several documented or proposed regulators, (2) provided a perspective on the nature of the influence being exerted by these regulators, (3) created a framework in which to perceive the relationships among these several regulators, and (4) presented a foundation for further exploring the mechanisms through which regulation is actually expressed.

Five sources of regulatory influence on the events and activities of amphibian forelimb regeneration have been reviewed. It should be apparent from the discussions contained therein that aspects of regulation from one source can overlap and even duplicate regulatory influence from another. Equally apparent should be the likely interactions among these sources of influence in achieving regulation of common physiological events. What might be more difficult to determine is whether the regulatory influence is of a permissive or an instructive nature, that is, whether it creates predisposing conditions that allow particular events of regeneration to occur or whether it actively causes these events by directing (or modulating) specific cellular functions. For example, does a cicatrix form, rather than a blastema, in hypophysectomized newts because the organism cannot meet the metabolic demands for regeneration (with significant proliferation and growth) yet must take measures to repair a severe wound? (permissive influence) Or are certain hormones (released by or controlled by the pituitary gland, for instance) required to trigger particular cellular events that will lead to the dedifferentiation of local stump tissues? (instructive influence) Discrimination of this nature is not essential to recognizing the significance of a regulator or to defining aspects of regeneration that are affected; however, it will be important to fully characterize the physiological mechanism through which the regulatory influence is expressed. Of equal importance will be to discriminate proximate from remote sources of influence. In this regard, nerves might be expected to provide a proximate, or direct, influence, while some hormones might be considered as remote regulators, insofar as they influence events of regeneration by affecting metabolism systemically and generally rather than locally. A need for such discrimination does arise where two or more regulators appear to be involved in affecting a single physiological change or event. Do these regulators accomplish their effects synergistically? Is there an interaction between them involving one regulator acting on the other, with the second regulator providing the direct influence on the event(s) regulated? For example, bioelectrical potentials, hormones, and nerves all appear to be important during the early stages of regeneration. Although manipulation of each might yield qualitatively equivalent results, the bioelectrical potentials and hormones might affect regenerative events by influencing the production and release of neurotrophic substances.

Our current understanding of the nature of regulation of amphibian forelimb regeneration does not yet make these issues a focus for investigation; however, we are rapidly approaching a time when such questions will begin to arise more frequently. A brief recapitulation and synthesis may be helpful to placing our perception of regulatory influences on regeneration into perspective. One convenient way to accomplish this is to focus our attention on the expression of regulatory influence on a phase-by-phase basis. Using Schmidt's (1968) designations (preblastemic, blastemic, morphogenetic) seems especially convenient in this regard. Table 9.1 and the following discussion attempt to summarize the major points of influence identified in the preceding chapters. (In the interest of brevity and clarity, references in this section are made to the chapters of this volume in which more detailed discussion, and specific reference citations, can be found.)

Influences on Events of the Preblastemic Phase

During the preblastemic phase of regeneration, wound healing and dedifferentiation are the predominant activities. All five sources of influence appear to be involved during this period (see Table 9.1). Interference with these regulators, either singly or in combination, typically results in abortive or delayed regeneration.

Wound Healing

During the wound healing period, leukocytes are the active agents causing demolition, removing debris, and promoting fibrogenesis (see Chap. 8). In addition, the wound epidermis can also perform a phagocytic function at this time (see Chap. 3). The degree of fibrogenesis (and/or fibroplasia) that occurs during this period is also regulated by bioelectrical, hormonal, and neural factors as evidenced by the altered patterns of wound healing observed following manipulation of Na^+-dependent currents (see Chap. 2), hypophysectomy or thyroidectomy (see Chap. 4), or denervation (see Chap. 5). It is not known whether the influence exerted in these latter instances is directed at local fibroblast and connective tissue populations, is mediated through modifications of leukocyte behaviors, or is expressed through some other means. Nevertheless, these sources of regulatory influence determine the degree of

TABLE 9.1. Regulatory Influences Exerted upon Major Events Occurring During Forelimb Regeneration

Events of limb regeneration		Sources of regulatory influence				
Phase of regeneration	Activity	Bioelectrical potentials (Chap. 2)	Integumentary system (Chaps. 3 and 7)	Endocrine system (Chaps. 4 and 7)	Nerves (Chaps. 5–7)	Blood cells/ immune system (chap. 8)
Preblastemic	Wound healing and debridement	Positive (?)[a]	Positive	Positive	Positive	Pos/neg (?)
	Dedifferentiation and accumulation of blastema	Positive (?)	Positive	Positive (?)	Positive	Pos/neg (?)
Blastemic	Proliferation		Positive	Positive	Positive	Pos/neg (?)
	Biosynthesis and growth in mass			Positive	Positive	
	Differentiation and early histogenesis		Negative	Positive	Pos/neg (?)	
Morphogenetic	Late histogenesis			Positive		
	Pattern formation	Positive (?)	Positive			Negative (?)

[a]For each source of regulatory influence discussed in this monograph, the general effect expressed on each of the events of regeneration listed above are noted. Positive means that the "regulator" identified acts to promote that particular event or activity. Negative means that the regulator acts to prevent or delay the expression of the event cited. Pos/neg denotes an effect that might either promote or adversely affect the specified activity. The (?) following a given designation signifies a proposed or suggested role for the regulator on the event in question. Where nothing appears, no influence on this event has been proposed or demonstrated, either positive or negative. A brief synopsis of the effects of these regulators on the events of regeneration in question is provided in the text. More detailed elaboration of these roles can be found in appropriate sections of the chapters cited under each source of regulatory influence.

fibroplasia and fibrogenesis that occurs during the wound healing phase, and through this ultimately determine whether or not the amputated limb will be regenerated.

Dedifferentiation and Blastema Accumulation

Dedifferentiation and the accumulation of a mass of embryonic-like mesenchymal cells, the regeneration blastema, then follow. The initial, outwardly positive bioelectrical potential measured at the wound surface begins to diminish during this period (see Chap. 2). What is actually occurring at this time remains poorly understood; however, it appears that the skin might participate in at least two ways (see Chap. 3). On the one hand, the dermis (and perhaps the epidermis as well) contributes cells to the forming blastemal mesenchyme population. This, however, is not a regulatory role. On the other hand, the apical epidermal cap that forms determines the orientation of the future outgrowth of the blastema, once formed (see Chap. 3). In addition, this apical epidermal cap appears to be active in maintaining the dedifferentiated state of the blastemal mesenchyme subjacent to it (see Chaps. 3 and 7). This is a role whose expression begins now but that is actually more significant during the later stages.

Nerves appear to play a role in the accumulation of the blastema, once dedifferentiation has occurred, by promoting proliferation of the mesenchymal cells (see Chaps. 5 and 6). However, nerves might also affect events of dedifferentiation itself, since the morphologies of the mesenchymal cells in denervated stumps do not resemble those of typical blastemal cells (see Chap. 5). The precise role of hormones on dedifferentiation and the accumulation of the blastema is even less clearly determined. Most attempts at endocrine manipulation (coincident with or before amputation) modify the wound healing response (see Chap. 4). Although some dedifferentiation is often reported, the failure of a blastema to form is usually attributed to ineffective or blocked interaction between the apical epidermal cap (or wound epidermis) and the subjacent soft tissues of the stump (see Chaps. 3 and 4).

A role for the immune system during the period of dedifferentiation remains speculative. Because of perceived parallels between the regeneration blastema and sarcomas, it has been suggested that mild immunostimulation should promote the formation and growth of the blastema (see Chap. 8). By extrapolation, one would presume that immunosuppression (or insufficient immunostimulation) and strong immunostimulation would delay or prevent the formation of a blastema; the former because of a failure to stimulate growth [proliferation (?)], and the latter because of immune rejection of dedifferentiated cells. However, none of the studies conducted thus far have documented any alteration in dedifferentiation or blastema accumulation because of modified immune status.

Influences Over Events of the Blastemic Phase

This period of regeneration is characterized by a high mitotic index, significant blastemal growth, and the initial expressions of differentiated phenotypes. Epidermal, neural, and endocrine influences (see Chaps. 3–7) are required for these events (see Table 9.1).

Proliferation

The wound epidermis (apical epidermal cap) (see Chaps. 3 and 7) and neurotrophic factors (see Chaps. 5–7) promote proliferation of the blastemal cells. The nature of the active substances (several candidates exist) and when and how they exert their influences remain to be determined, however (see Chap. 6). There appears to be positive hormonal (insulin) influence on proliferation as well (see Chap. 7). The immune system is postulated to affect blastemal growth, presumably through an effect on proliferation, at least in part (see Chap. 8). Retardation and arrest of blastemal growth has occurred following attempts at immunological manipulation; however, in the absence of demonstrated alterations in immunological status, a definite role for the immune system as a regulator of blastemal cell proliferation must be considered tentative.

Biosynthesis and Growth

Nerves (see Chaps. 5–7) and some hormones (see Chap. 7) have been shown to affect macromolecular synthesis by regeneration blastemas, both in vivo and in vitro. In the absence of an adequate nerve supply, DNA, RNA, and protein synthesis are reduced. Following administration of putative neurotrophic substances, synthesis of these macromolecules is increased. The identities of any of the substances synthesized and their specific relation to the physiological events of limb regeneration remain unknown however.

Differentiation

Certain hormones, for example, insulin, appear to promote events suggestive of differentiation (e.g., sulfate incorporation into glycosaminoglycans; see Chap. 7). There is thus some evidence for hormonal control of differentiation occurring during regeneration. Other manifestations will undoubtedly arise as more studies are undertaken with this objective in mind. In contrast, the apical epidermis appears to exert a negative influence on differentiation, that is, it favors the dedifferentiated state (see Chaps. 3 and 7). Nerves might also be viewed as negative regulators of differentiation in that their promotion of proliferation would appear to forestall the expression of any overt tendency toward a differentiated phenotype. However, their ability to affect the macromolecular synthesis of the blastema to a significant extent (up to 60%) raises the possibility of their effecting changes in phenotypic expression as well.

Influences on the Expression of Morphogenesis

As regeneration advances further, the blastemal cells express specific phenotypic traits, tissues are reorganized, and the pattern and form of the limb are restored. Of all of the sources of regulatory influence discussed, only the skin has been shown to clearly and dramatically affect pattern formation, and only the endocrine system has been shown to affect activities related to histogenesis (see Table 9.1).

Late Histogenesis

The influence of insulin on sulfate incorporation, cited above (see Chap. 7) suggests a role for the endocrine system during the early morphogenetic period, when histogenesis would be occurring. Instances in which defects in histogenesis have been observed following denervation have often been attributed to secondary consequences of reduced blastemal mass (too few cells). Influences on histogenesis have not been ascribed to either the integumentary or the immune systems.

Pattern Formation

Instances of defective form or pattern in regenerates have been reported following alterations in bioelectrical potentials (see Chap. 2) and following treatments with a variety of immunoactive compounds (see Chap. 8). In these instances, the investigators have raised the question about bioelectrical or immunological influences on pattern formation, but have refrained from proposing such an influence. Only in the case of the integument has a clear role in pattern formation been proposed and demonstrated. This influence appears to derive from two sources (see Chap. 3): The apical epidermal cap is one source, the dermis is the other. The apical epidermal cap serves to provide information about the external boundary of the regenerate, while the dermis plays the more major role, controlling the form and asymmetry of the regenerate.

Summary and Conclusion

As a newt proceeds to regenerate its limb following amputation, a complex series of events begins to unfold. These events provide opportunities to explore fundamental questions of developmental biology and cellular physiology. During this process, gene expression among cells is changed (dedifferentiation and redifferentiation), proliferation and growth occur, biosynthetic activities are modified, and the form and pattern of a highly complex structure are specified. In achieving the completion of regeneration, regulation of the above-mentioned activities is accomplished by several sources; three are well established (the skin, nerves, and the endocrine system), while two others are less universally recognized and accepted (bioelectrical potentials and the immune system). Our understanding of the roles played by these regulators—what events they regulate and how they regulate them—are far from complete. We have made considerable progress in determining which major events of regeneration are affected by several of these sources of regulatory influence (see Table 9.1). Substantially more work is required, however, in elucidating how these influences are expressed in finally achieving regeneration of an amputated forelimb, or other structure.

In brief, our general understanding of the regulation of amphibian limb regeneration appears to be as follows.

Bioelectrical potentials that are outwardly positive are associated with wounds (and developing limb buds). Manipulation of the currents producing these potentials can qualitatively modify the regenerative response. These potentials appear to coincide with the early stages of regeneration particularly; however, their precise relationship to the events of regeneration remains unclear.

The *skin* plays a variety of roles in regeneration. Early on, the epidermis can contribute to wound healing and phagocytosis, but later serves to maintain mesenchymal cells in the dedifferentiated state and provides information about the external boundary of the regenerate. The dermis appears to contribute to the blastemal mesenchyme, but more importantly is the major source of the skin's patterning information, specifying major axes of asymmetry and thus the form and pattern of the limb.

The *endocrine system* appears to play a role in all facets of regeneration, except pattern formation; however, there is disagreement about how active this role might be. On the one hand, it appears that certain hormones can specifically promote particular events in regeneration (at least in vitro), but there is also a sense that the endocrine influence might be chiefly permissive, that is, allowing events of regeneration to occur by providing an adequate metabolic background. Early attempts to discover a particular "regeneration hormone" have failed and more recent notions reflect a sense of the multi-faceted nature of hormone action (and interaction).

The *nervous system* also appears to influence all events of regeneration, except pattern formation. The evidence for its role in stimulating proliferation of blastema cells is especially strong; however, the nature of the active neurotrophic agents and their mechanism of action remain to be determined.

The *immune system* has only a proposed role in regeneration. There is some circumstantial evidence that argues for this role; however, it suffers from a lack of evidence for altered immunological status occurring during or affecting regeneration. Nevertheless, leukocytes and the immune system undoubtedly can affect the quality of the inflammatory response and, most likely, affect the initiation of regeneration.

We have begun to define those events of vertebrate limb regeneration that are affected by a variety of regulators (see Table 9.1). Although we do not yet know which specific cellular events are affected (e.g., Does the reduction in proliferation following denervation result from a G_1 or a G_2 block? Is the apparent suppression of regeneration following hypophysectomy the result of limited metabolic resources or of a failure to modify gene expression?), research on limb regeneration has reached the stage that allows questions of this nature to be asked. Moreover, refinements in experimental strategies (e.g., organ culture studies; see Chap. 7) promise to generate answers to several of these questions in the near future.

Regeneration of a mammalian limb is still many years into the future; however, we have made significant progress toward understanding the regulation of limb regeneration. As this understanding continues to increase, the information provided will also extend our understanding of the regulation of cellular and developmental physiology generally. Furthermore, since study of vertebrate limb regeneration requires analyses of regulation in situ (e.g., pattern formation, bioelectric influences) it thus generates a more comprehensive percpetion of the actual regulation of cellular and developmental activities.

References

Abercrombie, M. (1957). Localized formation of new tissue in an adult mammal. Symp. Soc. Exp. Biol. *11*:235–254.

Adair, L.-G., and C.-H. Taban (1971). Comparison of wound healing and/or regeneration in mammals and urodeles. Arch. Sci. (Geneva) *24*:499–516.

Andrews, E. J. (1976). Immunostimulation as a modulator of regeneration and growth. Am. Zool. *16*:260.

Andrews, W., and C. P. Hickman (1974). *Histology of the Vertebrates*, Mosby, St. Louis, pp. 44–50.

Anton, H. J. (1965). The origin of blastema cells and protein synthesis during forelimb regeneration in *Triturus*. In *Regeneration in Animals and Related Problems* (V. Kiortsis and H. A. L. Trampusch, eds.), North Holland, Amsterdam, pp. 377–395.

Asboe-Hansen, G. (1961). Endocrine control of connective tissue. In *Inflammation and Disease of Connective Tissues* (L. C. Mills and J. H. Moyer, eds.), Saunders, Philadelphia, pp. 38–43.

Asbury, A. K. (1975). The biology of Schwann cells. In *Peripheral Neuropathy*, Vol. 1, (P. J. Dyck, P. K. Thomas, and E. H. Lambert, eds.), Saunders, Philadelphia, pp. 201–212.

Atwell, W. J. (1921). The morphogenesis of the hypophysis in the tailed amphibia. Anat. Rec. *22*:373–390.

Babich, G. L., and J. E. Foret (1973). Effects of dibutyryl cyclic AMP and related compounds on newt limb regeneration blastemas *in vitro*. II. ^{14}C-Leucine incorporation. Oncology *28*:89–95.

Barker, A. T., L. F. Jaffe, and J. W. Vanable, Jr. (1982). The glabrous epidermis of cavies contains a powerful battery. Am. J. Physiol. *242*:R358–R366.

Bar-Maor, J. A., and G. Gitlin (1961). Attempted induction of forelimb regeneration by augmentation of nerve supply in young rat. Plast. Reconst. Surg. Trans. Bull. *27*:460–461.

Barnes, D., and G. Sato (1980). Methods of growth of cultured cells in serum-free medium. Anal. Biochem. *102*:255–270.

Bassina, J. A. (1940). An inquiry into regeneration by the method of calculating mitotic coefficients. Bull. Exp. Biol. Med. (USSR) *10*:389–391.

Becker, R. O. (1960). The bioelectric field pattern in the salamander and its simulation by an electronic analog. IRE Trans. Med. Electron. 7:202–207.

Becker, R. O. (1961a). The bioelectric factors in amphibian limb regeneration. J. Bone Joint Surg. *43A*:643–656.

Becker, R. O. (1961b). Search for evidence of axial current flow in peripheral nerves of salamander. Science *134*:101–102.

Becker, R. O. (1972). Stimulation of partial limb regeneration in rats. Nature (London) *235*:109–111.

Becker, R. O., and J. A. Spadaro (1972). Electrical stimulation of partial limb regeneration in mammals. Bull. N.Y. Acad. Med. *48*:627–641.

Benos, D. L., S. A. Simons, L. J. Mandel, and P. M. Cala (1976). Effect of amiloride and some of its analogues on cation transport in isolated frog skin and thin lipid membranes. J. Gen. Physiol. *68*:43–63.

Berelowitz, M., D. Le Roith, H. von Schenk, C. Newgard, M. Szabo, L. A. Frohman, J. Shiloach, and J. Roth (1982). Somatostatin-like immunoactivity and biological activity is present in *Tetrahymena pyriformis*, a ciliated protozoan. Endocrinology *110*:1939–1944.

Berman, R., H. A. Bern, C. S. Nicoll, and R. C. Strohman (1964). Growth-promoting effects of mammalian prolactin and growth hormone in tadpoles of *Rana catesbeiana*. J. Exp. Zool. *156*:353–360.

Bern, H. A., and C. S. Nicoll (1968). The comparative endocrinology of prolactin. Rec. Prog. Horm. Res. *24*:681–720.

Berridge, M. J. (1975). Control of cell division: A unifying hypothesis. J. Cyclic Nucleotide Res. *1*:305–320.

Betz, W. J., J. H. Caldwell, R. R. Ribchester, K. R. Robinson, and R. F. Stump (1980). Endogenous electric field around muscle fibres depends on the Na^+ and K^+ pump. Nature (London) *287*:235–237.

Bezkorovainy, A. (1980). *Biochemistry of Nonheme Iron*, Plenum, New York.

Bodemer, C. W. (1963). Distribution of ribonucleic acid in the regenerating urodele limb as determined by autoradiographic localization of uridine-H^3. Anat. Rec. *142*:457–468.

Bodemer, C. W., and N. B. Everett (1959). Localization of newly synthesized proteins in regenerating newt limbs as determined by radio autoradiographic localization of injected methionine-^{35}S. Dev. Biol. *1*:327–342.

Borgens, R. B. (1982a). What is the role of naturally produced electric current in vertebrate regeneration and healing? Int. Rev. Cytol. *76*:245–298.

Borgens, R. B. (1982b). Mice regrow the tips of their foretoes. Science *217*:747–750.

Borgens, R. B. (1983). The role of ionic current in the regeneration and development of the amphibian limb. In *Limb Development and Regeneration* (J. F. Fallon and A. I. Caplan, eds.), Alan R. Liss, New York, pp. 597–608.

Borgens, R. B. (1984a). Endogenous ionic currents traverse intact and damaged bone. Science *225*:478–482.

Borgens, R. B. (1948b). Are limb development and limb regeneration both initiated by an integumentary wounding? A hypothesis. Differentiation *28*:87–93.

Borgens, R. B., J. W. Vanable, Jr., and L. F. Jaffe (1977a). Bioelectricity and regeneration: Large currents leave the stumps of regenerating newt limbs. Proc. Nat. Acad. Sci. (U.S.A.) *74*:4528–4532.

Borgens, R. B., J. W., Vanable, Jr., and L. F. Jaffe (1977b). Bioelectricity and regeneration. I. Initiation of frog limb regeneration by minute currents. J. Exp. Zool. *200*:403–416.

Borgens, R. B., J. W. Vanable, Jr., and L. F. Jaffe (1979a). Small artificial currents enhance *Xenopus* limb regeneration. J. Exp. Zool. *207*:217–255.

Borgens, R. B., J. W. Vanable, Jr., and L. F. Jaffe (1979b). Bioelectricity and regeneration. Bioscience *29*:468–474.

Borgens, R. B., J. W. Vanable, Jr., and L. F. Jaffe (1979c). Reduction of sodium dependent stump currents disturbs urodele limb regeneration. J. Exp. Zool. *209*:377–386.

Borgens, R. B., J. W. Vanable, Jr., and L. F. Jaffe (1979d). Role of subdermal currents shunts in the failure of frogs to regenerate. J. Exp. Zool. *209*:49–55.

Borgens, R. B., L. F. Jaffe, and M. J. Cohen, (1980). Large and persistent electrical currents enter the transected lamprey spinal cord. Proc. Nat. Acad. Sci. (U.S.A.) 77:1208–1213.

Borgens, R. B., M. F. Rouleau, and L. E. DeLanney (1983). A steady efflux of ionic current predicts hind limb development in the Axolotl. J. Exp. Zool. *228*:491–503.

Borgens, R. B., M. E. McGinnis, J. W. Vanable, Jr., and E. S. Miles (1984). Stump currents in regenerating salamanders and newts. J. Exp. Zool. *231*:249–256.

Breedis, C. (1952). Induction of accessory limbs and of sarcoma in the newt (*Triturus viridescens*) with carcinogenic substances. Cancer Res. *12*:861–866.

Briggs, R., and T. J. King (1952). Transplantation of living nuclei from blastula cells into enucleated frogs' eggs. Proc. Nat. Acad. Sci. (U.S.A.) *38*:455–463.

Briggs, R., and T. J. King (1957) Changes in the nuclei of differentiating endodermal cells as revealed by nuclear transplantation. J. Morphol. *100*:269–312.

Brinkley, B. R., S. M. Cox, and D. A. Pepper (1980). Structure of the mitotic apparatus and chromosomes after hypotonic treatment of mammalian cells *in vitro*. Cytogenet. Cell Genet. *26*:165–174.

Brockes, J. P. (1984). Mitogenic growth factors and nerve dependence of limb regeneration. Science *225*:1280–1287.

Bromley, S. C. (1977). Accumulation of glucocorticoids in regeneration areas of limbs of the newt, *Notophthalmus viridescens*. J. Exp. Zool. *201*:101–108.

Bromley, S. C., and C. S. Thornton (1974). Effect of highly purified growth hormone on limb regeneration in the hypophysectomized newt, *Notophthalmus viridescens*. J. Exp. Zool. *190*:143–153.

Brownstein, M. J., E. A. Mroz, J. S. Kizer, M. Palkovits, and S. E. Leeman (1976). Regional distribution of substance P in the brain of the rat. Brain Res. *116*:299.

Brunst, V. V., and E. A. Cheremetieva (1935). Sur la perte local du pouvoir régénérateur chez le triton et l'axolotl par l'irradiation avec les rayons X. Arch. Zool. Exp. (Genève) *78*:57–67.

Bryant, S. V., and L. Iten (1976). Supernumerary limbs in amphibians: Experimental production in *Notophthalmus viridescens* and a new interpretation of their formation. Dev. Biol. *50*:212–234.

Bryant, S. V., D. Fyfe, and M. Singer (1971). The effects of denervation on the ultrastructure of young limb regenerates in the newt, *Triturus*. Dev. Biol. *24*:577–595.

Bryant, S. V., V. French, and P. J. Bryant (1981). Distal regeneration and symmetry. Science *212*:993–1002.

Bullough, W. S. (1950). Mitotic activity in the tissues of dead mice and in tissues kept in physiological salt solutions. Exp. Cell. Res. *1*:410–420.

Burnett, A. L., C. E. Kary, and A. M. Lagorio (1971). Induction of growth in newt and frog limbs after perfusion with extracts from newt blastemas. Nature (London) *234*:98–99.

Butler, E. G. (1933). The effects of X-radiation on the regeneration of the forelimb of *Amblystoma* larvae. J. Exp. Zool. *65*:271–316.

Butler, E. G. (1935). Studies on limb regeneration in X-rayed *Amblystoma* larvae. Anat. Rec. *62*:295–307.

Butler, E. G., and J. P. O'Brien (1942). Effects of localized x-irradiation on regeneration of the urodele limb. Anat. Rec. *84*:407–413.

Butler, E. G., and W. O. Puckett (1940). Studies on cellular interactions during limb regeneration in *Amblystoma* J. Exp. Zool. *84*:223–239.

Butler, E. G., and O. E. Schotté (1941). Histological alterations in denervated non-regenerating limbs of urodele larvae. J. Exp. Zool. *88*:307–341.

Butler, E. G., and M. B. Ward (1965). Reconstruction of the spinal cord following ablation in urodele larvae. J. Exp. Zool. *160*:47–66.

Caplan, A. I., and S. Koutroupas (1973). The control of muscle and cartilage development in the chick limb bud: The role of differential vascularization. J. Embryol. Exp. Morphol. *29*:571–583.

Carafoli, E., and M. Crompton (1976). Calcium ions and mitochondria. In *Calcium in Biological Systems* (C. J. Duncan, ed.), Cambridge University Press, London, pp. 89–115.

Carlone, R. L., and J. E. Foret (1979). Stimulation of mitosis in cultured limb blastemata of the newt, *Notophthalmus viridescens*. J. Exp. Zool. *210*:245–252.

Carlone, R. L., and M. P. Rathbone (1985). Partial purification of a low molecular weight growth factor from chicken brain. J. Cell. Physiol. (in press).

Carlone, R. L., M. Ganagarajah, and M. P. Rathbone (1981). Bovine pituitary fibroblast growth factor has neurotrophic activity for newt limb regenerates and skeletal muscle *in vitro*. Exp. Cell. Res. *132*:15–21.

Carlson, B. M. (1969). Inhibition of limb regeneration in the axolotl after treatment of the skin with actinomycin D. Anat. Rec. *16*:389–402.

Carlson, B. M. (1970). Relationship between the tissue and epimorphic regeneration of muscles. Am. Zool. *10*:175–186.

Carlson, B. M. (1974a). Factors controlling the initiation and cessation of early events in the regenerative process. In *Neoplasia and Cell Differentiation* (G. V. Sherbet, ed.), Karger, Basel, pp. 60–105.

Carlson, B. M. (1974b). Morphogenetic interactions between rotated skin cuffs and underlying stump tissues in regenerating axolotl forelimbs. Dev. Biol. *39*:263–285.

Carlson, B. M. (1975). The effects of rotation and positional change of stump tissues upon morphogenesis of the regenerating axolotl limb. Dev. Biol. *47*:269–291.

Carlson, B. M. (1978). Types of morphogenetic phenomena in vertebrate regenerating systems. Am. Zool. *18*:869–882.

Carrel, A. (1921). Cicatrization of wounds. XII. Factors initiating regeneration. J. Exp. Med. *34*:425–434.

Carrel, A. (1922). Growth-promoting function of leukocytes. J. Exp. Med. *36*:385–391.

Chafouleas, J. G., W. E. Bolton, H. Hidaka, A. E. Boyd III, and A. R. Means (1982). Calmodulin and the cell cycle. Cell *28*:41–50.

Chalkley, D. T. (1954). A quantitative histological analysis of forelimb regeneration in *Triturus viridescens*. J. Morphol. *94*:21–70.

Chalkley, D. T. (1959). The cellular basis of limb regeneration. In *Regeneration in Vertebrates* (C. S. Thornton, ed.), University of Chicago Press, Chicago, pp. 34–58.

Chang, M. M., and S. E. Leeman (1970). Isolation of sialogogic peptide from bovine hypothalamic tissue and its characterization as substance P. J. Biol. Chem. *245*:4784–4790.

Chang, M. M., S. E. Leeman, and H. D. Niall (1971). Amino acid sequence of substance P. Nature New Biol. *232*:86–87.

Chapron, C. (1974). Mise en évidence du rôle, dans la régénération des amphibiens, d'une glycoproteine sécrétée par la cape apicale: Étude cytochimique et autoradiographique en microscopie électronique. J. Embryol. Exp. Morphol. *32*:133–145.

Cheung, W. Y. (1980). Calmodulin plays a pivotal role in cellular regulation. Science *207*:19–27.

Choo, A. F., D. M. Logan, and M. P. Rathbone (1978). Nerve trophic effects: An *in vitro* assay for factors involved in regulation of protein synthesis in regenerating amphibian limbs J. Exp. Zool. *206*:347–354.

Choo, A. F., D. M. Logan, and M. P. Rathbone (1981). Nerve trophic effects: Partial purification from chick embryo brains of proteins that stimulate protein synthesis in cultured newt blastemata. Exp. Neurol. *73*:558–570.

Clarke, B. J. (1978). Restoration of regenerative ability of X-rayed newt limbs after grafting of proximal or distal skin. Growth *42*:275–295.

Cline, M. J. (1975). *The White Cell*, Harvard University Press, Cambridge.

Cohen, N. (1966). Tissue transplantation immunity in the adult newt, *Diemictylus viridescens*. II. The rejection phase: First- and second-set allograft rejections and lack of sexual dimorphism. J. Exp. Zool. *163*:173–190.

Conn, M. E., G.E. Dearlove, and M. H. Dresden (1979). Selection of a chemically defined medium for culturing adult newt forelimb regenerates. In Vitro *15*:409–414.

Connelly, T. G., R. A. Tassava, and C. S. Thornton (1968). Limb regeneration and survival of prolactin treated hypophysectomized adult newts. J. Morphol. *126*:365–371.

Copeland, D. E. (1943). Cytology of the pituitary gland in developing and adult *Triturus viridescens*. J. Morphol. *72*:379–409.

Cox, T. C., and R. H. Alvarado (1979). Electrical and transport characteristics of skin of larval *Rana catesbieana*. Am. J. Physiol. *237*:R74–R79.

Cuatrecasas, P. (1974). Membrane receptors. Annu. Rev. Biochem. *43*:169–214.

Dearlove, G. E., and D. L. Stocum (1974). Denervation-induced changes in soluble protein content during forelimb regeneration in the adult newt, *Notophthalmus viridescens*. J. Exp. Zool. *190*:317–327.

Deck, J. D. (1971). The effects of infused materials upon the regeneration of newt limbs. II. Extracts from newt brain and spinal cord. Acta Anat. *79*:321–332.

Deck, J. D., and J. N. Dent (1970). The effects of infused material upon the regeneration of newt limbs. III. Blastemal extracts and alkaline phosphatase in irradiated limb stumps. Anat. Rec. *168*:525–536.

Deck, J. D., and C. B. Futch (1969). The effects of infused materials upon the regeneration of newt limbs. I. Blastemal extracts on denervated limb stumps. Dev. Biol. *20*:322–348.

DeLustro, F., A. M., Machel, B. DeLustro, and E. C. LeRoy (1983). Human monocyte regulation of connective tissue growth. Am. Zool. *23*:213–220.

Dent, J. N. (1966). Maintenance of thyroid function in newts with transplanted pituitary glands. Gen. Comp. Endocrinol. *6*:401–408.

Desha, D. L. (1974). Irradiated cells and blastema formation in the adult newt, *Notophthalmus viridescens*. J. Embryol. Exp. Morphol. *32*:405–416.

Desselle, J.-C. (1968). Restauration de la régénération par des implantes de cartilage dans les membres irradiés de *Triturus cristatus*. C.R. Acad. Sci. (Paris) *267*:1642–1645.

Desselle, J.-C. and M. Goncharoff (1978). Cytophotometric detection of the participation of cartilage grafts in regeneration of X-rayed urodele limbs. Biol. Cell. *33*:45–54.

DeVries, G. H., L. N. Minier, and B. L. Lewis (1983). Further studies on the mitogenic response of cultured Schwann cells to rat CNS axolemma-enriched fractions. Dev. Brain Res. *9*:87–93.

Dodson, J. W., and E. D. Hay (1974). Secretion of collagenous stroma by isolated epithelium grown *in vitro*. Exp. Cell Res. *65*:215–220.

Donaldson, D. J., and J. M. Mason (1975). Cancer-related aspects of regeneration research: A review. Growth *39*:475–496.

Donaldson, D. J., and J. M. Mason (1977). Inhibition of epidermal cell migration by concanavalin A in skin wounds of the adult newt. J. Exp. Zool. *200*:55–64.

Donaldson, D. J., and J. M. Mason (1977). Inhibition of protein synthesis in newt epidermal cells: Effects on cell migration and concanavalin A-mediated inhibition of migration *in vivo*. Growth *42*:243–252.

Drachman, D. B. (1974). Trophic actions of the neuron: An introduction. Ann. N.Y. Acad. Sci. *228*:3–5.

Drachman, D. B., and M. Singer (1971). Regeneration in botulinum-poisoned forelimbs of the newt, *Triturus*. Exp. Neurol. *32*:1–11.

Dresden, M. H. (1969). Denervation effects on newt limb regeneration: DNA, RNA and protein synthesis. Dev. Biol. *19*:311–320.

Dresden, M. H., and J. Gross (1970). The collagenolytic enzyme of the regenerating limb of the newt, *Triturus viridescens*. Dev. Biol. *22*:129–137.

Droin, A. (1959). Potentialités morphogènes dans la peau de *Triton* en régénération. Rev. Suisse Zool. *66*:641–709.

DuBois-Reymond, E. (1843). Volaufiger Abrifs einer Untersuchung über den Sogenannten froschstrom und über die Electromotorischen fische. Ann. Phys. Chim. *58*:1–73.

DuBois-Reymond, E. (1860). *Untersuchungen über Tierische Elecktrizitat*, Vol. 2, Reimer, Berlin.

Dumont, A. E. (1965). Fibroplasia: A sequel to lymphocyte exudation. In *The Inflammatory Process* (B. W. Zweifach, L. Grant, and R. T. McClusky, eds.), Academic, New York, pp. 535–557.

Dunis, D. A., and M. Namenwirth (1977). The role of grafted skin in the regeneration of X-irradiated axolotl limbs. Dev. Biol. *56*:97–109.

Dunlap, M. K., and D. J. Donaldson (1978). Inability of colchicine to inhibit newt epidermal cell migration or prevent concanavalin A-mediated inhibition of migration: Studies *in vivo*. Exp. Cell Res. *116*:15–19.

Dunphy, J. E., and K. N. Udupa (1955). Chemical and histochemical sequence in the healing of wounds. N. Engl. J. Med. *253*:847–851.

Duprat, A. M., and P. Kan (1981). Stimulating effect of the divalent cation ionophore A23187 on *in vitro* neuroblast differentiation. Experientia *37*:154–157.

Eggert, R. C. (1966). The response of X-irradiated limbs of adult urodeles to autografts of normal cartilage. J. Exp. Zool. *161*:369–390.

Eisen, A. Z., and J. Gross (1965). The role of epithelium and mesenchyme in the production of a collagenolytic enzyme and a hyaluronidase in the anuran tadpole. Dev. Biol. *12*:408–418.

Erlij, D., and H. H. Ussing (1978). Transport across amphibian skin. In *Membrane Transport in Biology*, Vol. 3, Springer-Verlag, Berlin, pp. 175–208.

Erslev, A. J., and T. G. Gabuzda (1979). *Pathophysiology of Blood*, Saunders, Philadelphia, pp. 113–131.

Faber, J. (1960). An experimental analysis of regional organization in the regenerating forelimb of the axolotl (*Ambystoma mexicanum*). Arch. Biol. *71*:1–72.

Faber, J. (1971). Vertebrate limb ontogeny and limb regeneration: Morphogenetic parallels. Adv. Morphog. *9*:127–147.

Fini, M. E., and R. E. Sicard (1980). Limb regeneration of the adult newt (*Notophthalmus viridescens*) in the absence of the spleen. Wilhelm Roux Arch. *189*:77–79.

Foret, J. E., and G. L. Babich (1973a). Inhibition by acetylcholine of protein synthesis in cultured limb regenerates. Oncology *27*:281–285.

Foret, J. E., and G. L. Babich (1973b). Effects of dibutyryl cyclic AMP and related compounds on newt limb regeneration blastemas *in vitro*. I. ^{3}H-Thymidine incorporation. Oncology *28*:83–88.

Fox, W. F. (1982). Axial designations for the limbs of *Ambystoma* and *Notophthalmus*. Axolotl. Newsl. *11*:6–8.

Francis, E. T. B. (1934). *The Anatomy of the Salamander*, Clarendon Press, Oxford.

French, V., P. J. Bryant, and S. V. Bryant (1976). Pattern regulation in epimorphic fields. Science *193*:969–981.

Fritsch, C. (1911). Experimentelle studien über Regenerationsvorgänger des Gleidmassenskeletts. Zool. Jahrb. Abt. Physiol. *30*:377–472.

Garling, D. G., and R. A. Tassava (1984). Injury, nerve, and wound epidermis related electrophoretic and fluorographic protein patterns in forelimbs of adult newts. J. Exp. Zool. *231*:221–241.

Gelfant, S. (1977). A new concept of tissue and tumor cell proliferation. Cancer Res. *37*:3845–3862.

Genuth, S., and H. E. Lebovitz (1965). Stimulation of insulin release by corticotropin. Endocrinology *70*:1093–1099.

Geraudie, J., and M. Singer (1978). Nerve dependent macromolecular synthesis in the epidermis and blastema of the adult newt regenerate. J. Exp. Zool. *203*:455–460.

Gesner, B. M. (1965). The "life history" and function of lymphocytes. In *The Inflammatory Process* (B. W. Zweifach, L. Grant, and R. T. McClusky, eds.), Academic, New York, pp. 281–322.

Glade, R. W. (1963). Effects of tail skin, epidermis, and dermis on limb regeneration in *Triturus viridescens* and *Siredon mexicanum*. J. Exp. Zool. *152*:169–194.

Globus, M. (1970). *In Vitro* Studies of Innervated Tail and Limb Regenerates in Urodeles: The Influence of Nerves on Growth and Differentiation. Ph.D. thesis, University of Toronto, Toronto, Canada.

Globus, M. (1978). Neurotrophic contribution to a proposed tripartite control of the mitotic cycle in the regeneration blastema of the newt (*Notophthalmus* (*Triturus*) *viridescens*). Am. Zool. *18*:855–868.

Globus, M., and R. A. Liversage (1975a). Differentiation *in vitro* of innervated tail regenerates in larval *Ambystoma*. J. Embryol. Exp. Morphol. *33*:803–812.

Globus, M., and R. A. Liversage (1975b). *In vitro* studies of limb regeneration in adult *Diemictylus viridescens*: Neural dependence of blastema cells for growth and differentiation. J. Embryol. Exp. Morphol. *33*:813–829.

Globus, M., and S. Vethamany-Globus (1976). An *in vitro* analogue of early chick limb bud outgrowth. Differentiation *6*:91–96.

Globus, M., and S. Vethamany-Globus (1977). Transfilter mitogenic effect of dorsal root ganglia on cultured regeneration blastemata, in the newt, *Notophthalmus viridescens*. Dev. Biol. *56*:316–328.

Globus, M., S. Vethamany-Globus, and Y. C. I. Lee (1980). Effect of apical epidermal cap on mitotic cycle and cartilage differentiation in regeneration blastemata in the newt, *Notophthalmus viridescens*. Dev. Biol. *75*:358–372.

Globus, M., S. Vethamany-Globus, A. Kesik, and G. Milton (1983). Roles of neural peptide substance P and calcium in blastema cell proliferation in the newt, *Notophthalmus viridescens*. In *Limb Development and Regeneration* (J. F. Fallon and A. I. Caplan, eds.), Alan R. Liss, New York, pp. 513–524.

Godlewski, E. (1928). Untersuchungen über Auslosung und Hemmung der Regeneration beim Axolotl. Wilhem Roux Arch. *114*:108–143.

Goldfine, I. D., R. A. Vigneri, A. L. Jones, G. J. Smith, and K. Y. Wong. (1977). Localization of the nuclear binding site for insulin. Clin. Res. *25*:160A.

Golub, E. S. (1981). Immune tolerance and autoreactivity: Do they point to a regulatory role of the immune system for normal cell function. Cell *27*:417–418.

Gospodarowicz, D. (1974). Localisation of a fibroblast growth factor and its effect alone and with hydrocortisone on 3T3 cell growth. Nature (London) *249*:123–127.

Gospodarowicz, D. (1983). Growth factors and their action *in vivo* and *in vitro*. J. Pathol. *141*:201–233.

Gospodarowicz, D., and A. L. Mescher (1977). A comparison of the responses of cultured fibroblasts and chondrocytes to fibroblast and epidermal growth factors. J. Cell Physiol. *93*:117–128.

Gospodarowicz, D., and A. L. Mescher (1980). Fibroblast growth factor and the control of vertebrate regeneration and repair. Ann. N.Y. Acad. Sci. *339*:151–174.

Gospodarowicz, D., and J.S. Moran (1976). Growth factors in mammalian cell culture. Annu. Rev. Biochem. *45*:531–558.

Gospodarowicz, D., P. Rudland, J. Lindstrom, and K. Benirschke (1975). Fibroblast growth factor: Its localization, purification, mode of action, and physiological significance. Adv. Metab. Disord. *8*:301–335.

Gospodarowicz, D., J. S. Moran, and A. L. Mescher (1978a). Cellular specificities of fibroblast growth factor and epidermal growth factor. In *Molecular Control of Proliferation and Differentiation* (J. Papaconstantinou and W. J. Rutter, eds.), Academic, New York, pp. 33–64.

Gospodarowicz, D., H. Bialecki, and G. Greenburg (1978b). Purification of the fibroblast growth factor activity from bovine brain. J. Biol. Chem. *223*:3736–3743.

Gospodarowicz, D., N. Savion, L. Giguere, and J. P. Tauber (1981). Effects of fibroblast growth factor and basal lamina on the growth of normal diploid cells in serum-free medium. In *Physiological Peptides and New Trends in Radioimmunology* (C. A. Bizallon, ed.), Elsevier/North Holland, Amsterdam, pp. 127–147.

Gospodarowicz, D., G. M., Lui, and J. Cheng (1982). Purification in high yield of brain fibroblast growth factor by preparative isoelectric point focusing at pH 9.6. J. Biol. Chem. *257*:12,266–12,276.

Gospodarowicz, D., J. Cheng, and M. Lirette (1983). Bovine brain and pituitary fibroblast growth factors: Comparison of their abilities to support the proliferation of human and bovine vascular endothelial cells. J. Cell. Biol. 97:1677–1685.

Gospodarowicz, D., J. Cheng, G. M. Lui, A. Baird, and P. Böhlent (1984). Isolation by heparin–sepharose affinity chromatography of brain fibroblast growth factor: Identity with pituitary fibroblast growth factor. Proc. Nat. Acad. Sci. (U.S.A.) *81*:6963–6967.

Goss, R. J. (1956a). Regenerative inhibition following limb amputation and immediate insertion into the body cavity. Anat. Rec. *126*:15–27.

Goss, R. J. (1956b). The regenerative responses of amputated limbs to delayed insertion into the body cavity. Anat. Rec. *126*:283–297.

Goss, R. J. (1957a). The relation of skin to defect regulation in regenerating half limbs. J. Morphol. *100*:547–563.

Goss. R. J. (1957b). The effect of partial irradiation on the morphogenesis of limb regenerates. J. Morphol. *101*:131–148.

Goss, R. J. (1961). Regeneration of vertebrate appendages. Adv. Morphol. *1*:103–152.

Goss, R. J. (1969). *Principles of Regeneration*, Academic, New York.

Goss, R. J. (1972). Wound healing and antler regeneration. In *Epidermal Wound Healing* (H. I. Maibach and D. T. Rovee, eds.), Year Book Medical, Chicago, pp. 219–228.

Graem, N., and K. Helweg-Larsen (1979). Mitotic activity and delay in fixation of tumor tissue. Acta Pathol. Microbiol. Scand. Sect. A *87*:375–378.

Grillo, R. S. (1971). Changes in mitotic activity during limb regeneration in *Triturus*. Oncology *25*:347–355.

Grillo, H. C., C. M. LaPiere, H. H. Dresden, and J. Gross (1968). Collagenolytic activity in regenerating forelimbs of the adult newt (*Triturus viridescens*). Dev. Biol. *17*:571–583.

Grimes, L. N., and R. J. Goss (1970). Regeneration of holes in rabbit ears. Am. Zool. *10*:537.

Gulati, A. K., A. A. Zalewski, and A. H. Reddi (1983). An immunofluorescent study of the distribution of fibronectin and laminin during limb regeneration in the adult newt. Dev. Biol. *96*:355–365.

Gurdon, J. B. (1974). *The Control of Gene Expression in Animal Development*. Harvard University Press, Cambridge, p. 15.

Guth, L. (1968). "Trophic" influences of nerve on muscle. Physiol. Rev. *48*:645–687.

Guth, L. (1971). Degeneration and regeneration of tastebuds. In *Handbook of Sensory Physiology*, Vol. 14, Springer-Verlag, Berlin, pp. 63–74.

Gutmann, E. (1976). Neurotrophic relations. Annu. Rev. Physiol. *38*:177–216.

Haist, R. E. (1959). Islet cell function. Ann. N. Y. Acad. Sci. *82*:266–286.

Hall, A. B., and O. E. Schotté (1951). Effects of hypophysectomies upon the initiation of regenerative processes in the limbs of *Triturus viridescens*. J. Exp. Zool. *118*:363–387.

Hasegawa, T., K. Saito, I. Kimura, and E. Ozawa (1981). Fe^{3+} promotes *in vitro* growth of myoblasts and other cells from chick embryos. Proc. Jpn. Acad. *57*:206–210.

Hay, E. D. (1952). The role of epithelium in amphibian limb regeneration, studies by haploid and triploid transplants. Am. J. Anat. *91*:447–481.

Hay, E. D. (1956). Effects of thyroxine on limb regeneration in the newt, *Triturus viridescens*. Bull. Johns Hopkins Hosp. *99*:262–285.

Hay, E. D. (1966a). Metabolic patterns in limb development and regeneration. In *Organogenesis* (R. L. DeHaan and H. Ursprung, eds.), Holt, Rinehart and Winston, New York, pp. 315–336.

Hay, E. D. (1966b.) *Regeneration*, Holt, Rinehart, and Winston, New York.
Hay, E. D. (ed.) (1981). *Cell Biology of Extracellular Matrix*, Plenum, New York.
Hay, E. D., and D. A. Fischman (1961). Origin of the blastema in regenerating limbs of the newt *Triturus viridescens*. An autoradiographic study using tritiated thymidine to follow cell proliferation and migration. Dev. Biol. *3*:26–59.
Hellmich, W. (1930). Untersuchungen über Herkunft und Determination des Regenerativen Materiels bei Amphibien. Wilhelm Roux Arch. *121*:135–202.
Hellmich, W. (1931). Histology of regeneration in different species of adult and larval urodeles. Anat. Rec. *48*:303–307.
Herlitzka, A. (1910). Ein Beitrag zur Physiologie der Generation. Wilhelm Roux Arch. *10*:126–158.
Hessler, A. C. (1974). A Re-evaluation of Requirements for Limb Regeneration in the Newt, *Triturus viridescens*. Ph.D. thesis, University of Vermont, Burlington, Vermont.
Hessler, A. C., and R. Landesman (1981a). An investigation of the prolactin–thyroxine synergism in newt limb regeneration. J. Morphol. *167*:103–108.
Hessler, A. C., and R. Landesman (1981b). Hormone dependent changes in the apical connective tissue during early stages of forelimb regeneration in the hypophysectomized newt, *Notophthalmus viridescens*. J. Morphol. *168*:297–308.
Hightower, J. A., and R. L. St. Pierre (1971). Hemopoietic tissues of the adult newt, *Notophthalmus viridescens*. J. Morphol. *135*:299–308.
Hinchliffe, J. R., and Johnson, D. R. (1980). *The Development of the Vertebrate Limb*, Clarendon, Oxford.
Hinterberger, T. J., and J. A. Cameron (1983). Muscle and cartilage differentiation in axolotl limb regeneration blastema cultures. J. Exp. Zool. *226*:399–407.
Hirsch, J. G. (1965). Neutrophil and eosinophil leukocytes. In *The Inflammatory Process* (B. W. Zweifach, L. Grant, and R. T. McClusky, eds.), Academic, New York, pp. 245–280.
Hoffman, C. W., and J. N. Dent (1977a). Hormonal effects on mitotic rhythm in the epidermis of the red-spotted newt. Gen. Comp. Endocrinol. *32*:512–521.
Hoffman, C. W., and J. N. Dent (1977b). Hormonal regulations of cellular proliferation in the epidermis of the red-spotted newt. Gen. Comp. Endocrinol. *32*:522–530.
Hoffman, R. S. (1940). The growth-activating effect of extracts of adult and embryonic tissues of rat on fibroblast colonies *in vitro*. Growth *4*:361–376.
Hoffman, D. K., D. Kleinebeckel, and W. Luther. (1978). Limb regeneration from X-irradiated tails of *Ambystoma mexicanum* following transplantation of flank skin from region adjacent to hindlimb. Wilhelm Roux Arch. *185*:227–234.
Hokfelt, T., J. O. Kellerth, G. Nilsson, and B. Pernow (1975). Experimental immunohistochemical studies on the localization and distribution of substance P in cat primary sensory neurons. Brain Res. *100*:235–252.
Holder, N., and R. Glade (1984). Skin glands in the axolotl: The creation and maintenance of a spacing pattern. J. Embryol. Exp. Morphol. 79:97–112.
Holtzer, H. (1958). Lack of organ-determining role of skin during regeneration. Anat. Rec. *132*:455.
Horbay, G. L. A., and R. A. Liversage (1977). *In vitro* regeneration of the spinal cord in *Ambystoma* larvae. Am. Zool. *17*:881A.
Ide-Rozas, A. (1937). Die cytologischen Verhältnisse bei der Regeneration von Kaulquappenextremitäten. Wilhelm Roux Arch. *135*:552–608.
Ii, I., I. Kimura, T. Hasegawa, and E. Ozawa (1981). Transferrin is an essential component of chick embryo extract for avian myogenic cell growth *in vitro*. Proc. Jpn. Acad. *57*:211–216.
Ii, I., I. Kimura, and E. Ozawa (1982). A myotrophic protein from chick embryo extract: Its purification, identity to transferrin, and indispensibility for avian myogenesis. Dev. Biol. *94*:366–377.

Illingworth, C. M. (1974). Trapped fingers and amputated fingertips in children. J. Pediatr. Surg. 9:853–858.

Illingworth, C. M., and A. T. Barker (1978). Measurement of electrical currents emerging during the regeneration of amputated finger tips in children. Clin. Phys. Physiol. Meas. *1*:87–89.

Iten, L., and S. V. Bryant (1973). Forelimb regeneration from different levels of amputation in the newt, *N. viridescens*: Length, rate, and stages. Wilhelm Roux Arch. *173*:263–282.

Iten, L. E., and S. V. Bryant (1975). The interaction between the blastema and stump in the establishment of the anterior–posterior and proximal–distal organization of the limb regenerate. Dev. Biol. *44*:119–147.

Jabaily, J., and M. Singer (1978). Neurotrophic and hepatotrophic stimulation of proliferation of embryonic chick muscle cells *in vitro*: Assay and partial characterization of mitogenic activity in chick embryonic organ and tissue extracts. Dev. Biol. *64*:189–202.

Jaffe, L. F. (1977). Electrophoresis along cell membranes. Nature (London) *265*:600–602.

Jaffe, L. F. (1980). Calcium explosions as triggers of development. Ann. N. Y. Acad. Sci. *339*:86–101.

Jaffe, L. F., and R. Nuccitelli (1974). An ultrasensitive vibrating probe for measuring steady extracellular current. J. Cell Biol. *63*:614–628.

Jaffe, L. F., and R. Nuccitelli (1977). Electrical controls of development. Annu. Rev. Biophys. Bioeng. *6*:445–476.

Jaffe, L. F., K. R. Robinson, and R. Nuccitelli (1974). Local cation entry and self-electrophoresis as an intracellular localization mechanism. Ann. N. Y. Acad. Sci. *238*:372–389.

Jennings, T., R. D. Jones, and A. Lipton (1979). A growth factor from spinal cord. J. Cell. Physiol. *100*:273–278.

Jessel, T. M. (1981). The role of substance P in sensory transmission and pain perception. In *Neurosecretion and Brain Peptides* (J. B. Martin, S. Reichlin, and K. L. Brick, eds.), Raven Press, New York, pp. 189–198.

Jessel, T. M., R. E. Siegel, and G. D. Fischbach (1979). Induction of acetylcholine receptors on cultured skeletal muscle by a factor extracted from brain and spinal cord. Proc. Nat. Acad. Sci. (U.S.A.) *76*:5397–5401.

Johnson, M. C., and A. J. Schmidt (1974). Collagen synthesis in the regenerating forelimb of the adult newt, *Diemictylus viridescens*. J. Exp. Zool. *190*:185–198.

Kaltenbach, J. C., and N. B. Clark (1965). Direct action of thyroxine analogues on moulting in the adult newt. Gen. Comp. Endocrinol. *5*:74–86.

Kambara, S. (1953). The effect of thyroid and thiourea on regeneration of the tail in the newt, *Triturus pyrrhogaster*. Annot. Zool. Jpn. *26*:208–212.

Kamrin, A. A., and M. Singer (1959). The growth influence of spinal ganglia implanted into the denervated forelimb regenerate of the newt, *Triturus*. J. Morphol. *104*:415–439.

Karsenti, E., and P. Gounon (1979). Lampbrush chromosomes: Are they ATP dependent contractile structures? Biol. Cell. *34*:91–97.

Kauffman, S. A., R. M. Shymko, and K. Trabert (1978). Control of sequential compartment formation in *Drosophila*. Science *199*:259–270.

Kelly, D. J., and R. A. Tassava (1973). Cell division and ribonucleic acid synthesis during the initiation of limb regeneration in larval axolotls (*Ambystoma mexicanum*). J. Exp. Zool. *185*:45–54.

Kemp, N. E. (1961). Replacement of the larval basement lamella by adult-type basement membrane in anuran skin during metamorphosis. Dev. Biol. *3*:391–410.

Kimura, I., T. Hasegawa, T. Miura, and E. Ozawa (1981). Muscle trophic factor is identical to transferrin. Proc. Jpn. Acad. *57*:200–205.

Kintner, C. R., and J. P. Brockes (1984). Monoclonal antibodies identify blastemal cells derived from dedifferentiating muscle in newt limb regeneration. Nature (London) *308*:67–69.

Kirschner, L. B. (1973). Electrolyte transport across the body surface of freshwater fish and amphibia. In *Transport Mechanisms in Epithelia* (H. H. Ussing and N. A. Thorn, eds.), Munksgaard, Copenhagen, pp. 447–460.

Kissebah, A. H., and B. R. Tulloch (1976). A hypothesis for insulin action. In *Peptide Hormones* (J. A. Parsons, ed.), Macmillan, London, pp. 373–376.

Korneluk, R. G., and R. A. Liversage (1977). Distribution and blood serum level determinations of thyroid hormone during forelimb regeneration in adult *Notophthalmus viridescens*. Am. Zool. *17*:880A.

Korneluk, R. G., and R. A. Liversage (1984). Tissue regeneration in the amputated forelimb of *Xenopus laevis* froglets. Can. J. Zool. *62*:2383–2391.

Korneluk, R. G., M.-J. Anderson, and R. A. Liversage (1982). Stage dependency of forelimb regeneration on nerves in postmetamorphic froglets of *Xenopus laevis*. J. Exp. Zool. *220*:331–342.

Krieger, D. T. (1978). Hypothalamus. In *The Year in Endocrinology 1977* (S. H. Ingbar, ed.) Plenum, New York, pp. 1–26.

Landesman, R., and A. C. Hessler (1981). Limb regeneration following the discontinuation of growth hormone therapy in the hypophysectomized newt, *Notophthalmus viridescens*. J. Morphol. *168*:309–319.

Larsen, L. O. (1976). Physiology of molting. In *Physiology of the Amphibia*, Vol. 3 (B. Lofts, ed.), Academic, New York, pp. 54–100.

Lassalle, B. (1974a). Characteristiques des potentiels de surface des membres du triton *Pleurodeles watlii*, Michah. C. R. Acad. Sci. (Paris) Ser. D. *278*:483–486.

Lassalle, B. (1974b). Origine epidermique des potentiels de surface du membre de triton *Pleurodeles watlii*, Michah. C. R. Acad. Sci. (Paris) Ser. D. *278*:1055–1058.

Lassalle, B. (1979). Surface potentials and the control of amphibian regeneration. J. Embryol. Exp. Morphol. *53*:213–223.

Lassalle, B. (1980). Are surface potentials necessary for amphibian regeneration? Dev. Biol. *75*:460–466.

Lassalle, B. (1983). DNA synthesis and cellular proliferation in cultured newt blastemas. In *Limb Development and Regeneration* (J. F. Fallon and A. J. Caplan, eds.), Alan R. Liss, New York, pp. 547–555.

Laux, D. L. (1980). Determination and Analysis of the Cell Cycle Parameters of Three Tissues from the Regenerating Forelimb of the Newt. M.Sc. thesis, Ohio State University, Columbus, Ohio.

Lazard, L. (1968). Inhibiton de la régénération des membres par les rayons X, chez l'axolotl (*Ambystoma mexicanum*). Ann. Embryol. Morphol. *1*:49–59.

Lazarus, S. S., and B. W. Volk (1962). *The Pancreas in Human and Experimental Diabetes*, Grune and Stratton, New York, pp. 76–150.

Lebowitz, P., and M. Singer (1970). Neurotrophic control of protein synthesis in the regenerating limb of the newt, *Triturus*. Nature (London) *225*:824–827.

Leeman, S. E., and R. Hammerschlag (1967). Stimulation of salivary secretion by a factor extracted from hypothalamic tissues. Endocrinology *81*:803.

Lemke, G. E., and J. P. Brockes (1984). Identification and purification of glial growth factor. J. Neurosci. *4*:75–83.

Lentz, T. L. (1972). Distribution of leucine-^{3}H during axoplasmic transport within regenerating neurons as determined by electron-microscope radioautography. J. Cell Biol. *52*:719–732.

Lentz, T. L., J. S. Addis, and J. Chester (1981). Partial purification and characterization of a nerve trophic factor regulating muscle acetylcholinesterase activty. Exp. Neurol. *73*:542–557.

Le Roith, D., A. S. Liotta, J. Roth, J. Shiloach, M. E. Lewis, C. B. Pert, and D. T. Krieger

(1982). Corticotropin and β-endorphin-like materials are native to unicellular organisms. Proc. Nat. Acad. Sci. (U.S.A.) *79*:2086–2090.

Le Roith, D., J. Shiloach, M. Berelowitz, L. A. Frohman, A. S. Liotta, D. T. Krieger, and J. Roth (1983). Are messenger molecules in microbes the ancestors of the vertebrate hormones and tissue factors? Fed. Proc. *42*:2602–2607.

Levin, R. M., and B. Weiss (1979). Selective binding of antipsychotics and other psychoactive agents to the calcium-dependent activator of cyclic nucleotide phosphodiesterase. Pharmacol. Exp. Ther. *208*:454–459.

Lheureux, E. (1972). Contribution à l'étude du rôle de la peau et des tissues axiaux du membre dans le declenchement de morphogénèses régénératrices anormales chez le triton *Pleurodeles waltlii*, Michah. Ann. Embryol. Morphol. *5*:165–178.

Lheureux, E. (1975a). Nouvelles données sur les rôles de la peau et des tissues internes dans la régénération du membre du triton *Pleurodeles waltlii*, Michah (amphibian urodele). Wilhelm Roux. Arch. *176*:285–301.

Lheureux, E. (1975b). Régénération des membres irradiés de *Pleurodeles waltlii* Michah (urodèle), influence des qualitiés et orientations des greffons non-irradiés. Wilhelm Roux. Arch. *176*:303–327.

Li, C. H. (1956). Pituitary growth hormone as a metabolic hormone. Science *123*:617–619.

Libbin, R. M., P. Person, S. Papierman, D. Shah, D. Nevid, and H. Grob (1979). Partial regeneration of the above elbow amputated rat forelimb II. Electrical and mechanical facilitation. J. Morphol. *159*:439–451.

Liebman, E. (1946). On trephocytes and trephocytosis: A study on the role of leukocytes in nutrition and growth. Growth *10*:291–330.

Liebman, E. (1949). The leukocytes in regenerating limbs of *Triturus viridescens*. Growth *13*:103–118.

Lim, R. (1980). Glia maturation factor. Curr. Top. Dev. Biol. *16*:305–322.

Linsenmayer, T. F., and G. N. Smith, Jr. (1976). The biosynthesis of cartilage type collagen during limb regeneration in the larval salamander. Dev. Biol. *52*:19–30.

Litwiller, R. (1939). Mitotic index and size in regenerating amphibian limbs. J. Exp. Zool. *82*:273–286.

Liversage, R. A. (1959). The relation of the central and autonomic nervous systems to the regeneration of limbs in adult newts. J. Exp. Zool. *141*:75–117.

Liversage, R. A. (1962). Regeneration of homoplastically deplanted forelimbs following spinal cord ablation in *Amblystoma opacum* larvae. J. Exp. Zool. *151*:1–15.

Liversage, R. A. (1967). Hypophysectomy and forelimb regeneration in *Ambystoma opacum* larvae. J. Exp. Zool. *165*:57–70.

Liversage, R. A. (1973). Hypophysectomy and pectoral fin regeneration in adult *Fundulus heteroclitus* (killifish). Can. J. Zool. *51*:1047–1054.

Liversage, R. A., and B. J. Banks (1980). Insulin levels in the serum of intact and regenerating adult newts (*Notophthalmus viridescens*). J. Exp. Zool. *211*:259–266.

Liversage, R. A., and P. J. Brandes (1982). Forelimb regeneration in thyroidectomized adult newts following organ culture and autografting of the thyroid glands. Wilhelm Roux Arch. *191*:149–158.

Liversage, R. A., and K. R. S. Fisher (1974). Regeneration of the forelimb in adult hypophysectomized *Notophthalmus* (*Diemictylus*) *viridescens* given embryonic or adult chicken anterior pituitary extract. J. Exp. Zool. *190*:133–141.

Liversage, R. A., and M. Globus (1977). *In vitro* regeneration of innervated forelimb deplants of *Ambystoma* larvae. Can. J. Zool. *55*:1195–1199.

Liversage, R. A., and R. G. Korneluk (1978). Serum levels of thyroid hormone during forelimb regeneration in the adult newt, *Notophthalmus viridescens*. J. Exp. Zool. *206*:223–227.

Liversage, R. A., and L. Liivamagi (1971). Forelimb regeneration in hypophysectomized adult *Diemictylus viridescens* following organ culture and autoplastic implantation of the adenohypophysis. J. Embryol. Exp. Morphol. *26*:443–458.

Liversage, R. A., and D. S. McLaughlin (1983). Effects of delayed amputation on denervated forelimbs of adult newt. J. Embryol. Exp. Morphol. *75*:1–10.

Liversage, R. A., and B. W. Price (1973). Adrenocorticosteroid levels in adult *Diemictylus viridescens* plasma following hypophysectomy and forelimb amputation. J. Exp. Zool. *185*:259–264.

Liversage, R. A., and S. R. Scadding (1969). Re-establishment of forelimb regeneration in adult hypophysectomized *Diemictylus* (*Triturus*) *viridescens* given frog anterior pituitary extract. J. Exp. Zool. *170*:381–396.

Liversage, R. A., B. W. Price, W. C. Clarke and D. G. Butler (1971). Plasma adrenocorticosteriod levels in adult *Fundulus heteroclitus* (killifish) following hypophysectomy and pectoral fin amputation. J. Exp. Zool. *178*:23–28.

Liversage, R. A., M. P. Rathbone, and H. M. G. McLaughlin (1977). Changes in cyclic GMP levels during forelimb regeneration in adult *Notophthalmus viridescens*. J. Exp. Zool. *200*:169–175.

Liversage, R. A., W. E. Stewart, and D. S. McLaughlin (1984). *In vitro* studies of the influence of prolactin on tail regeneration in the adult newt (*Notophthalmus viridescens*). Wilhelm Roux Arch. *193*:379–387.

Lostroh, A. J., and P. Woodward (1958). Changes in the adrenal of the hypophysectomized C3H mouse with α-corticotropin and growth hormone. Endocrinology *62*:498.

Loyd, R. M., and T. C. Connelly (1981). Microdensitometric analysis of denervation effects on newt limb blastema cells. Experientia *37*:967–969.

Loyd, R. M., and R. A. Tassava (1980). DNA synthesis and mitosis in adult newt limbs following amputation and insertion into the body cavity. J. Exp. Zool. *214*:61–69.

Luchasen, J. R., J. G. White, and J. H. Kersey (1974). Mitogenic properties of a calcium ionophore, A23187. Proc. Nat. Acad. Sci. (U.S.A.) *71*:5088–5090.

McCaig, C. D., and K. R. Robinson (1982). The ontogeny of the transepidermal potential difference in frog embryos. Dev. Biol. *90*:335–339.

McConnell, I. (1975). Structure and function of lymphoid tissue. In *The Immune System: A Course on the Molecular and Cellular Basis of Immunity* (M. J. Hobart and I. McConnell, eds.), Blackwell Scientific, Oxford, pp. 197–205.

McCullough, W. D., and R. A. Tassava (1976). Determination of the blastema cell cycle in regenerating limbs of the larval axolotl, *Ambystoma mexicanum*. Ohio J. Sci. *76*:63–65.

Maciag, T., G. A. Hoover, and R. Weinstein (1982). High and low molecular weight forms of endothelial cell growth factor. J. Biol. Chem. *257*:5333–5336.

McLaughlin, H. M. G., and R. A. Liversage (1977). Preferential uptake of ^{125}I-thyroid hormone ^{3}H-thymidine in adult *Notophthalmus viridescens* tail and spinal cord regenerates. Am. Zool. *17*:880A.

McLaughlin, H. M. G., R. A. Liversage, M. P. Rathbone, and D. S. McLaughlin (1982). Cyclic AMP and cyclic GMP levels in regenerating adult newt forelimbs following anaesthetization. Can. J. Zool. *60*:976–983.

McLaughlin, H. M. G., M. P. Rathbone, R. A. Liversage, and D. S. McLaughlin (1983). Levels of cyclic GMP and cyclic AMP in regenerating forelimbs of adult newts following denervation. J. Exp. Zool. *225*:175–185.

McMahon, D. (1974). Chemical messengers in development: A hypothesis. Science *185*:1012–1021.

Maden, M. (1977). The regeneration of positional information in the amphibian limb. J. Theor. Biol. *69*:735–753.

Maden, M. (1978). Neurotrophic control of the cell cycle during amphibian limb regeneration. J. Embryol. Exp. Morphol. *48*:169–175.

Maden, M. (1979a). Neurotrophic and X-ray blocks in the blastema cell cycle. J. Embryol. Exp. Morphol. *50*:169–173.
Maden, M. (1979b). Regulation and limb regeneration: The effect of partial irradiation. J. Embryol. Exp. Morphol. *52*:183–192.
Maden, M. (1980). Intercalary regeneration in the amphibian limb and the law of distal transformation. J. Embryol. Exp. Morphol. *56*:201–209.
Maden, M. (1981). Morphallaxis in an epimorphic system: Size, growth control and pattern formation during amphibian limb regeneration. J. Embryol. Exp. Morphol. *65*:151–167.
Maden, M. (1982). Vitamin A and pattern formation in the regenerating limbs. Nature (London) *295*:672–675.
Maden, M., and K. Mustafa (1982). Axial organization of the regenerating limb: Asymmetrical behavior following skin transplantation. J. Embryol. Exp. Morphol. *70*:197–213.
Maden, M., and H. Wallace (1975). The origin of limb regenerates from cartilage grafts. Acta Embryol. Exp. *2*:77–86.
Maden, M., and H. Wallace (1976). How X-rays inhibit amphibian limb regeneration. J. Exp. Zool. *197*:105–114.
Mailman, M. L., and M. H. Dresden (1976). Collagen metabolism in the regenerating forelimb of *Notophthalmus viridescens*: Synthesis, accumulation, and maturation. Dev. Biol. *50*:378–394.
Margolis, R. L., and L. Wilson (1979). Regulation of the microtubule steady state *in vitro* by ATP. Cell *18*:673–679.
Markelonis, G. J., and T. H. Oh (1978). A protein fraction from peripheral nerve having neurotrophic effects on skeletal muscle cells in culture. Exp. Neurol. *58*:285–295.
Markelonis, G. J., V. F. Kemerer, and T. H. Oh (1980). Sciatin: Purification and characterization of a myotrophic protein from chicken sciatic nerves. J. Biol. Chem. *255*:8967–8970.
Markelonis, G. J., R. A. Bradshaw, T. H. Oh, J. L. Johnson, and O. J. Bates (1982). Sciatin is a transferrin-like polypeptide. J. Neurochem. *39*:315–320.
Marsh, G., and H. W. Beams (1952). Electrical control of morphogenesis in regenerating *Dugesia tigrina.* J. Cell. Comp. Physiol. *39*:191–213.
Martin, C. R. (1976). *Textbook of Endocrine Physiology*, Oxford University Press, New York.
Means, A. R., and J. R. Dedman (1980). Calmodulin—An intracellular calcium receptor. Nature (London) *285*:73–77.
Mescher, A. L. (1976). Effects on adult newt limb regeneration of partial and complete skin flaps over the amputation surface. J. Exp. Zool. *195*:117–127.
Mescher, A. L. (1982). Neurotrophic control of events in injured forelimbs of larval urodeles. J. Embryol. Exp. Morphol. *69*:183–192.
Mescher, A. L. (1983). Growth factors from nerves and their roles during limb regeneration. In *Limb Development and Regeneration* (J. F. Fallon and A. I. Caplan, eds.), Alan R. Liss, New York, pp. 501–512.
Mescher, A. L. (1984). Injury to nerves and the initiation of amphibian limb regeneration. Am. J. Anat. *169*:273–284.
Mescher, A. L., and D. Gospodarowicz (1979). Mitogenic effect of a growth factor derived from myelin on denervated regenerates of newt forelimbs. J. Exp. Zool. *207*:497–503.
Mescher, A. L., and J. J. Loh (1981). Newt forelimb regeneration blastemas *in vitro*: Cellular response to explantation and effects of various growth-promoting substances. J. Exp. Zool. *216*:235–245.
Mescher, A. L., and S. I. Munaim (1984). "Trophic" effect of transferrin on amphibian limb regeneration blastemas. J. Exp. Zool. *230*:485–490.

Mescher, A. L., and R. A. Tassava (1975). Denervation effects on DNA replication and mitosis during the initiation of limb regeneration in adult newts. Dev. Biol. *44*:187–197.

Michael, M., and J. Faber (1961). The self-differentiation of the paddle-shaped limb regenerate, transplanted with normal and reversed proximo-distal orientation after removal of the digital plate (*Ambystoma mexicanum*). Arch. Biol. *72*:301–329.

Miles, E. S., and J. W. Vanable, Jr. (1982). The effect of reducing stump currents on amphibian limb regeneration. Am. Zool. *22*:919.

Mitchison, J. M. (1971). *The Biology of the Cell Cycle*, Cambridge University Press, Cambridge.

Mizell, M. (1968). Limb regeneration: Induction in the newborn opossum. Science *161*:283–286.

Monroy, A. (1941). Ricerche sulle correnti elettriche dalla superficie del corpo di tritoni adulti normali e durante la rigenerazione degli arti e della coda. Pubbl. Stn. Zool. Napoli *18*:265–281.

Morgan, E. H. (1981). Transferrin biochemistry, physiology, and clinical significance. Mol. Aspects Med. *4*:1–123.

Morgan, T. H. (1901). *Regeneration*, Macmillan, New York.

Morzlock, F. V., and D. L. Stocum (1972). Neural control of RNA synthesis in regenerating limbs of the adult newt, *Triturus viridescens*. Wilhelm Roux Arch. *171*:170–180.

Murphy, P. (1976). *The Neutrophil*, Plenum, New York.

Nagle, B. W., and J. C. Egrie (1981). Calmodulin and ATPases in the mitotic apparatus. In *Mitosis/Cytokinesis* (A. M. Zimmerman and A. Forer, eds), Academic, New York, p. 337.

Namenwirth, M. (1974). The inheritance of cell differentiation during limb regeneration in the axolotl. Dev. Biol. *41*:42–56.

Nathanson, M. A., and E. D. Hay (1980a). Analysis of cartilage differentiation from skeletal muscle grown on bone matrix. I. Ultrastructural aspects. Dev. Biol. *78*:301–331.

Nathanson, M. A., and E. D. Hay (1980b). Analysis of cartilage differentiation from skeletal muscle grown on bone matrix. II. Chondroitin sulfate synthesis and reaction to exogenous glycosaminoglycans. Dev. Biol. *78*:332–351.

Needham, A. E. (1952). *Regeneration and Wound-healing*, Methuen, London.

Neufeld, D. A. (1980). Partial blastema formation after amputation in adult mice. J. Exp. Zool. *213*: 31–36.

Newman, S. A., and H. L. Frisch (1979). Dynamics of skeletal pattern formation in developing chick limb. Science *205*:662–668.

Nielsen, R., and R. W. S. Tomilson (1970). The effect of amiloride on sodium transport in the normal and moulting frog skin. Acta Physiol. Scand. *79*:238–243.

Niwelinski, J. (1958). The effect of prolactin and somatotrophin on the regeneration of the forelimb of the newt *Triturus alpestris* Laur. Folia Biol. (Krakow) *6*:9–36.

Norman, W. P., and A. J. Schmidt (1967). The fine structure of tissues in the amputated, regenerating limb of the adult newt, *Diemictylus*. J. Morphol. *123*:271–312.

Nouza, K., P. Nemec, and P. Draber (1973). The influence of host splenectomy on the tumor allograft-promoting effectiveness of antilymphocyte serum. Folia Biol. (Prague) *19*:267–272.

Ochs, S. (1974). Systems of material transport in nerve fibers (axoplasmic transport) related to nerve function and trophic control. Ann. N.Y. Acad. Sci. *228*:202–223.

Odland, G., and R. Ross (1968). Human wound repair: I, epidermal regeneration. J. Cell Biol. *39*:135–151.

Oh, T. H., and G. J. Markelonis (1978). Neurotrophic protein regulates muscle acetylcholinesterase in culture. Science *200*:337–339.

Oh, T. H., and G. J. Markelonis (1980). Dependence of *in vitro* myogenesis on a trophic protein present in chick embryo extract. Proc. Nat. Acad. Sci. (U.S.A.) 77:6922–6925.

Oh, T. H., and G. J. Markelonis (1982). Chicken serum transferrin duplicates the myotrophic effects of sciatin on cultured muscle cells. J. Neurosci. Res. *8*:535–545.

Oh, T. H., C. A. Sofia, Y. C. Kim, C. Carroll, H. H. Kim, G. J. Markelonis, and P. J. Reier (1981). Sciatin: Immunocytochemical localization of a myotrophic protein in chicken neural tissues. J. Histochem. Cytochem. *29*:1205–1212.

Olsen, C. L. (1980). Cellular Responses to Denervation and Re-innervation in Amputated Forelimbs of Larval *Ambystoma maculatum.* M.Sc. thesis, Ohio State University, Columbus, Ohio.

Olsen, C. L. (1982). Neurotrophic Influences on Cycling, Loss, and Rescue of Cells in Denervated and Re-innervated Forelimbs of *Ambystoma* Larvae. Ph.D. dissertation, Ohio State University, Columbus, Ohio.

Olsen, C. L., and R. A. Tassava (1984). Cell cycle and histological effects of reinnervation in denervated forelimb stumps of larval *Ambystoma.* J. Exp. Zool. *229*:247–258.

Olsen, C. L., P. M. Barger, and R. A. Tassava (1984). Rescue of blocked cells by reinnervation in denervated forelimb stumps of larval *Ambystoma.* Dev. Biol. (in press).

O'Steen, W. K., and B. E. Walker (1961). Radioautographic studies of regeneration in the common newt. II. Regeneration of the forelimb. Anat. Rec. *139*:547–556.

Overton, J. (1976). Scanning microscopy of collagen in the basement lamella of normal and regenerating frog tadpoles. J. Morphol. *150*:805–824.

Patrick, J., and R. Briggs (1964). Fate of cartilage cells in limb regeneration in the axolotl (*Ambystoma mexicanum*). Experientia *20*:431–432.

Peadon, A. M., and M. Singer (1966). The blood vessels of the regenerating limb of the adult newt *Triturus.* J. Morphol. *118*:79–90.

Pescitelli, M. J., Jr., and D. L. Stocum (1980). The origin of skeletal structures during intercalary regeneration of larval *Ambystoma* limbs. Dev. Biol. 79:255–275.

Petrosky, N. S., R. A. Tassava, and C. L. Olsen (1980). Cellular events in denervated limb stumps of *Ambystoma* larvae during re-innervation and subsequent regeneration. Experientia *36*:601–604.

Pickel, V. M., D. J. Reis, and S. E. Leeman (1977). Ultrastructural localization of substance P in neurons of rat spinal cord. Brain Res. *122*:534–540.

Pietrzyk-Walknowska, J. (1959). Regeneration of the forelimbs of *Xenopus laevis* inserted into the body cavity. Folia Biol. (Krakow) 7:1–8.

Polezhaev, L. V. (1972). *Loss and Restoration of Regenerative Capacity in Tissues and Organs of Animals*, Harvard University Press, Cambridge, Mass.

Polezhaev, L. V. (1979). Morphogenetic potencies of the regeneration blastema. Usp. Sovrem. Biol. *87*:287–303.

Polezhaev, L. V., and W. N. Faworina (1935). Über die Rolle des Epithels in den Anfanglichen Entwicklungsstadien einer Regenerationsanlage der Extremität beim Axolotl. Wilhelm Roux Arch. *133*:701–727.

Poo, M-M. (1981). In situ electrophoresis of membrane components. Annu. Rev. Biophys. Bioeng. *10*:245–276.

Poo, M.-M., and K. R. Robinson (1977). Electrophoresis of concanvalin A receptors along embryonic muscle cell membrane. Nature (London) *265*:602–605.

Poo, M.-M., W.-J. Poo, and J. W. Lam (1978). Lateral electrophoresis and diffusion of concanavalin A receptors in the membrane of embryonic muscle cell. J. Cell Biol. *76*:483–501.

Popiela, H. (1973). The Ultrastructure of Regenerating and Denervated Limbs of the Larval Salamander, *A. mexicanum.* M.Sc. thesis, Ohio State University, Columbus, Ohio.

Popiela, H., and S. Ellis (1981). Neurotrophic factor: Characterization and partial purification. Dev. Biol. *83*:266–277.

Popiela, H., S. Ellis, and B. W. Festoff (1982). Dose-dependent initiation of myogenesis by neurotrophic factor. J. Neurosci. Res. *8*:547–567.

Popiela, H., D. Taylor, S. Ellis, R. Beach, and B. Festoff (1984). Regulation of mitotic activity and the cell cycle in primary chick muscle cells by neurotransferrin. J. Cell. Physiol. *119*:234–240.

Powell, D., S. E. Leeman, G. W. Tregear, H. D. Niall, J. T. Potts, Jr. (1973). Radio immuno-assay for substance P. Nature (London) *241*:252–254.

Powell, J. A. (1969). Analysis of histogenesis and regenerative ability of denervated forelimb regenerates of *Triturus viridescens*. J. Exp. Zool. *170*:125–148.

Prehn, R. T. (1970). Immunosurveillance, regeneration, and oncogenesis. Prog. Exp. Tumor Res. *14*:1–24.

Prehn, R. T. (1971). Perspectives on oncogenesis: Does immunity stimulate or inhibit neoplasia? J. Reticuloendothel. Soc. *10*:1–16.

Prehn, R. T. (1972). The immune reaction as a stimulator of tumor growth. Science *176*:170–171.

Prehn, R. T. (1980). Neoplasia. In *Principles of Pathobiology* (R. B. Hill and M. F. LaVia, eds.), Oxford University Press, New York, pp. 200–254.

Prehn, R. T., and M. Lappé (1971). An immunostimulation theory of tumor development. Transplant Rev. 7:26–54.

Rahmani, T. M. Z. (1960). Conflix de potentialitiés morphogènes et duplicature. Rev. Suisse Zool. 67:589–675.

Rasmussen, H. (1975). Ions as "second messengers." In *Call Membranes* (G. Weissman and R. Claiborne, eds.), HP Publishing, New York, pp. 203–212.

Rathbone, M. P., J. Petri, A. F. Choo, D. M. Logan, R. L. Carlone, and J. E. Foret (1980). Noradrenaline and cyclic AMP independent growth stimulation in newt limb blastemata. Nature (London) *283*:387–388.

Ratner, N., L. Glaser, and R. P. Bunge (1984). PC12 cells as a source of neurite-derived cell surface mitogen, which stimulates Schwann cell division. J. Cell Biol. *98*:1150–1155.

Reddi, N. C., and K. E. Kuettner (1981). Vascular invasion of cartilage: Correlation of morphology with lysozyme, glycosaminoglycans, protease, and protease-inhibitory activity during endochrondral bone development. Dev. Biol. *82*:217–223.

Reiter, R. S., and M. Solursh (1982). Mitogenic property of the apical ectodermal ridge. Dev. Biol. *93*:28–35.

Repesh, L. A., and L. T. Furcht (1982). Distribution of fibronectin in regenerating limbs of the adult newt, *Notophthalmus viridescens*. Differentiation *22*:125–131.

Repesh, L. A., and J. C. Oberpriller (1978). Scanning electron microscopy of epidermal cell migration in wound healing during limb regeneration in the adult newt, *Notophthalmus viridescens*. Am. J. Anat. *151*:539–556.

Repesh, L. A., and J. C. Oberpriller (1980). Ultrastructural studies on migrating epidermal cells during the wound healing stage of regeneration in the adult newt, *Notophthalmus viridescens*. Am. J. Anat. *159*:187–208.

Repesh, L. A., L. T. Furcht, and D. Smith (1981). Immunocytochemical localization of fibronectin in limb tissues of the adult newt, *Notophthalmus viridescens*. J. Histochem. Cytochem. 8:937–945.

Reyer, R. W. (1954). Regeneration of the lens in the amphibian eye. Q. Rev. Biol. *29*:1–46.

Richardson, D. (1940). Thyroid and pituitary hormones in relation to regeneration. I. The effect of anterior pituitary hormone on regeneration of the hind leg in normal and thyroidectomized newts. J. Exp. Zool. *83*:407–429.

Richardson, D. (1945). II. Regeneration of the hind leg of the newt, *Triturus viridescens* with different combinations of thyroid and pituitary hormones. J. Exp. Zool. *100*:417–429.

Riddiford, L. M. (1960). Autoradiographic studies of tritiated thymidine infused into the blastema of the early regenerate in the adult newt, *Triturus*. J. Exp. Zool. *144*:25–32.

Robinson, K. R. (1983). Endogenous electrical current leaves the limb and prelimb region of the *Xenopus* embryo. Dev. Biol. 97:203–211.

Robison, G.A., R. W. Butcher, and E. W. Sutherland (1971). *Cyclic AMP*, Academic, New York.

Rollman-Dinsmore, C., and S. V. Bryant (1982). Pattern regulation between hind- and forelimb after blastema exchanges and skin grafts in *Notophthalmus viridescens*. J. Exp. Zool. *223*:51–56.

Rose, F. C., and S. M. Rose (1965). The role of normal epidermis in recovery of regenerative ability in X-rayed limbs of *Triturus*. Growth *29*:361–393.

Rose, S. M. (1942). A method for inducing limb regeneration in adult Anura. Soc. Exp. Biol. Med. *49*:408–410.

Rose, S. M. (1944). Methods of initiating limb regeneration in adult Anura. J. Exp. Zool. *95*:149–170.

Rose, S. M. (1945). The effect of NaCl in stimulating regeneration of limbs of frogs. J. Morphol. 77:119–139.

Rose, S. M. (1948). Epidermal dedifferentiation during blastema formation in regenerating limbs of *Triturus viridescens*. J. Exp. Zool. *108*:337–361.

Rose, S.M. (1970). *Regeneration: Key to Understanding Normal and Abnormal Growth and Development*, Appleton-Century-Crofts, New York.

Rose, S. M. and F. C. Rose (1974). Electrical studies on normally regenerating on x-rayed, and on denervated limb stumps of *Triturus*. Growth *38*:363–380.

Rosenberg, M., D. Lewinson, and M. R. Warburg (1982). Ultrastructural studies of the epidermal leydig cell in larvae of *Salamandra salamandra* (Caudata, Salamandrida). J. Morphol. *174*:275–281.

Rosenzweig, J. L., J. Havrankova, M. A. Lesniak, M. Brownstein, and J. Roth (1980). Insulin is ubiquitous in extrapancreatic tissues of rats and humans. Proc. Nat. Acad. Sci. (U.S.A.) 77:572–576.

Ross, R. (1968). The fibroblast and wound repair. Biol. Rev. *43*:51–96.

Ross, R. (1970). Wound-healing, a review. In *Chemistry and Molecular Biology of the Intercellular Matrix*, Vol. 3 (E. A. Belazs, ed.), Academic, New York, pp. 1739–1751.

Roth, J., D. Le Roith, J. Shiloach, J. L. Rosenzweig, M. A. Lesniak, and J. Havrankova (1982). The evolutionary origin of hormones, neurotransmitters and other extracellular chemical messengers. N. Engl. J. Med. *306*:523–527.

Rothstein, H., J. J. VanWyk, J. H. Hayden, S. R. Gordon, and A. Weinsieder (1980). Somatomedin C: Restoration *in vivo* of cycle traverse in G_0/G_1 blocked cells of hypophysectomized animals. Science *208*:410–412.

Ruben, L. (1983). IgM memory: Long-lived hapten-specific memory in the newt, *Notophthalmus viridescens*. Immunology *48*:385–392.

Ruben, L. N. and B. F. Edwards (1975). The inference of lymphoid cell heterogeneity in the newt, *Triturus viridescens*, from hapten-carrier antigen binding studies. In *Phylogeny of Thymus and Bone Marrow-Bursa Cells* (R. K. Wright and E. L. Cooper, eds.), Elsevier/North Holland, Amsterdam, pp. 161–168.

Ruben, L. N., and E. Selker (1975). Polyfunctional antigen-binding specificity in hapten-carrier responses of the newt, *Triturus viridescens*. In *Immunological Phylogeny* (W. H. Hildemann and A. A. Benedict, eds.), Plenum, New York, pp. 387–395.

Rzehak, K., and M. Singer (1966). Limb regeneration and nerve fiber number in *Rana sylvatica* and *Xenopus laevis*. J. Exp. Zool. *162*:15–22.

Salley, J. D., and R. A. Tassava (1981). Responses of denervated adult newt limb stumps to re-innervation and reinjury. J. Exp. Zool. *215*:183–189.

Salpeter, M. (1965). Disposition of nerve fibers in regenerating limbs of the adult newt *Triturus*. J. Morphol. *117*:201–212.

Salpeter, M. M., and M. Singer (1959). The fine structure of the adepidermal reticulum in the basal membrane of the skin of the newt, *Triturus*. J. Biophys. Biochem. Cytol. *6*:35–40.

Salpeter, M. M., and M. Singer (1960a). Differentiation of the submicroscopic adepidermal membrane during limb regeneration in adult *Triturus*, including a note on the use of the term basement membrane. Anat. Rec. *136*:27–40.

Salpeter, M., and M. Singer (1960b). The fine structure of mesenchymatous cells in the regenerating forelimb of the adult newt. Dev. Biol. *2*:516–534.

Salzer, J. L., and R. P. Bunge (1980). Studies of Schwann cell proliferation. I. An analysis in tissue culture of proliferation during development, Wallerian degeneration, and direct injury. J. Cell. Biol. *84*:739–752.

Salzer, J. L., R. P. Bunge, and L. Glaser (1980a). Studies of Schwann cell proliferation. III. Evidence for the surface localization of the neurite mitogen. J. Cell Biol. *84*:767–778.

Salzer, J. L., A. K. Williams, L. Glaser, and R. P. Bunge (1980b). Studies of Schwann cell proliferation. II. Characterization of the stimulation and specificity of the response to a neurite membrane fraction. J. Cell Biol. *84*:753–766.

Sandor, T., and A. Z. Mehdi (1982). The evolutionary origin of extracellular chemical messengers. N. Engl. J. Med. *307*:629.

Sato, N. L., and S. Inoue (1973). Effects of growth hormone and nutrients on limb regeneration in hypophysectomized adult newts. J. Morphol. *140*:477–486.

Scadding, S. R. (1984). Forelimbs of the newt *Notophthalmus viridescens* continuously denervated for up to six weeks prior to amputation do not regenerate. J. Exp. Zool. *229*:155–161.

Scadding, S. R., and R. A. Liversage (1979). Axoplasmic transport in regenerating limbs of *Ambystoma maculatum* larvae. Experientia *35*:266–267.

Schackman, R. W., E. M. Eddy, and M. Shapiro, (1978). The acrosome reaction of *Strongylocentrotus purpuratus* sperm. Dev. Biol. *65*: 483–495.

Schaller, H. C. (1979). Neuropeptides in hydra. Trends Neurosci. *2*:120–122.

Schauble, M. K., and M. R. Nentwig (1974). Temperature and prolactin as control factors in newt forelimb regeneration. J. Exp. Zool. *187*:335–344.

Scher, C. D., R. C. Shepard, H. N. Antoniades, and C. D. Stiles (1979). Platelet-derived growth factor and the regulation of the mammalian fibroblast cell cycle. Biochim. Biophys. Acta *560*:217–241.

Schlaepfer, W. W., and R. P. Bunge (1973). Effects of calcium ion concentration on the degeneration of amputated axons in tissue culture. J. Cell Biol. *59*:456–470.

Schlaepfer, W. W., and M. B. Hasler (1979). Characterization of the calcium-induced disruption of neurofilaments in rat peripheral nerve. Brain Res. *168*:299–309.

Schmidt, A. J. (1958a). Forelimb regeneration of thyroidectomized adult newts. I. Morphology. J. Exp. Zool. *137*:197–226.

Schmidt, A. J. (1958b). Forelimb regeneration of thyroidectomized adult newt. II. Histology. J. Exp. Zool. *139*:95–135.

Schmidt, A. J. (1958c). Thyroxine effects on forelimb regeneration in the hypothyroid adult newt, *Triturus viridescens*. Anat. Rec. *132*:503–504A.

Schmidt, A. J. (1968). *Cellular Biology of Vertebrate Regeneration and Repair*, University of Chicago Press, Chicago.

Schotté, O. E. (1926a). Hypophysectomie et régénération chez les bactraciens urodèles. C. R. Seances Soc. Phys. Hist. Nat. (Genève) *43*:67–72.
Schotté, O. E. (1926b). Système nerveux et régénération chez le triton. Action globale des nerfs. Rev. Suisse Zool. *33*:1–211.
Schotté, O. E. (1953). The role of the pituitary, of ACTH, and of some adrenocortical extracts in the regeneration of limbs in adult *Triturus*. Anat. Rec. *117*:575–576.
Schotté, O. E. (1961). Systemic factors in initiation of regenerative processes in limbs of larval and adult amphibians. In *Synthesis of Molecular and Cellular Structures* (D. Rudnick, ed.), Ronald Press, New York, pp. 161–192.
Schotté, O. E., and R. H. Bierman (1956). Effects of cortisone and allied adrenal steroids upon limb regeneration in hypophysectomized *Triturus viridescens*. Rev. Suisse Zool. *63*:353–375.
Schotté, O. E., and E. G. Butler (1941). Morphological effects of denervation and amputation of limbs in urodele larvae. J. Exp. Zool. *87*:279–322.
Schotté, O. E., and E. G. Butler (1944). Phases in regeneration of urodele limbs and dependence on the nervous system. J. Exp. Zool. *97*:95–121.
Schotté, O. E., and J. L. Chamberlain (1955). Effects of ACTH upon limb regeneration in normal and in hypophysectomized *Triturus viridescens*. Rev. Suisse Zool. *62*:253–279.
Schotté, O. E., and W. B. Christiansen (1957). Influence of Amphenone B—a cortical inhibitor—upon regeneration in normal and in hypophysectomized newts. Anat. Rec. *128*:619–620.
Schotté, O. E., and A. Droin (1965). The competence of pituitaries and limb regeneration during metamorphosis of *Triturus* (*Diemictylus*) *viridescens*. Rev. Suisse Zool. *72*:205–223.
Schotté, O. E., and A. B. Hall (1952). Effect of hypophysectomy upon phases of regeneration in progress (*Triturus viridescens*). J. Exp. Zool. *121*:521–559.
Schotté, O. E., and M. Harland (1943). Amputation level and regeneration in limbs of late *Rana clamitans* tadpoles. J. Morphol. *73*:329–363.
Schotté, O. E., and S. R. Hilfer (1957). Initiation of regeneration in regenerates after hypophysectomy in adult *Triturus viridescens*. J. Morphol. *101*:25–55.
Schotté, O. E., and D. A. B. Lindberg (1954). Effect of xenoplastic adrenal transplants upon limb regeneration in normal and hypophysectomized newts (*Triturus viridescens*). Proc. Soc. Exp. Biol. Med. *87*:26–29.
Schotté, O. E., and R.A. Liversage (1959). Effects of denervation and amputation upon the initiation of regeneration in regenerates of *Triturus*. J. Morphol. *105*:495–528.
Schotté, O. E., and R. E. Sicard (1982). Cyclophosphamide-induced leukopenia and suppression of limb regeneration in the adult newt, *Notophthalmus viridescens*. J. Exp. Zool. *222*:199–202.
Schotté, O. E., and L. B. Smith (1961). Effects of ACTH and cortisone upon amputational wound healing processes in mice digits. J. Exp. Zool. *146*:209–229.
Schotté, O. E., and A. Tallon (1960). The importance of autoplastically transplanted pituitaries for survival and for regeneration of adult *Triturus*. Experientia *16*:72–74.
Schotté, O. E., and J. F. Wilber (1958). Effects of adrenal transplants upon forelimb regeneration in normal and in hypophysectomized adult frogs. J. Embryol. Exp. Morphol. *6*:247–269.
Schwabe, C., D. Le Roith, R. P. Thompson, J. Shiloach, and J. Roth (1983). Relaxin extracted from protozoa (*Tetrahymena pyriformis*). J. Biol. Chem. *258*:2778–2781.
Seichert, V., and Z. Rychter (1971). Vascularization of the developing anterior limb of the chick embryo. Folia Morphol. (Prague) *9*:367–377.
Seilern-Aspang, F., and K. Kratochwil (1965). Relation between regeneration and

tumor growth. In *Regeneration in Animals and Related Problems* (V. Kiortsis and H. A. L. Trampusch, eds.), North Holland, Amsterdam, pp. 452–472.
Selye, H. (1950). *The Physiology and Pathology of Exposure to Stress*, ACTA, Montreal.
Settles, H. E. (1978). Supernumerary regeneration caused by ninety degree turning of the limb skin in adult *Notophthalmus*. Growth *42*:297–307.
Sicard, R. E. (1981a). The effects of putative immunological manipulations upon the rate of limb regeneration in adult newts, *Notophthalmus viridescens*. IRCS Med. Sci. *9*:692–693.
Sicard, R. E. (1981b). Morphologically aberrant limb regeneration in adult newts induced by putative immunological manipulations. IRCS Med. Sci. *9*:694–695.
Sicard, R. E. (1983a). Blood cells and their role in amphibian limb regeneration. I. Changes in circulating blood cell counts during forelimb regeneration. Exp. Cell Biol. *51*:51–569.
Sicard, R. E. (1983b). Blood cells and their role in amphibian limb regeneration. II. Effects of putative immunological manipulations on circulating blood cell counts during regeneration. Exp. Cell Biol. *51*:109–114.
Sicard, R. E. (1983c). Neurotrophic influence on proliferation–differentiation decisions during amphibian forelimb regeneration: An hypothesis. BioSystems *16*:65–73.
Sicard, R., and A. F. DiNicola (1974). Retardation of progressive forelimb regeneration by treatment with neuropharmacological agents. Oncology *30*:442–447.
Sicard, R. E., and W. T. Laffond (1983). Putative immunological influence upon amphibian forelimb regeneration. I. Effects of several immunoactive agents on regeneration rate and gross morphology. Exp. Cell Biol. *51*:337–344.
Sidman, R. L., and M. Singer (1960). Limb regeneration without innervation of the apical epidermis in the adult newt, *Triturus*. J. Exp. Zool. *144*:105–110.
Simpson, D. M., and R. Ross (1972). The neutrophilic leukocyte in wound repair: A study with antineutrophil serum. J. Clin Invest. *51*:2009–2023.
Singer, M. (1942). The nervous system and regeneration of the forelimb of adult *Triturus*. I. The role of the sympathetics. J. Exp. Zool. *90*:377–399.
Singer, M. (1943). The nervous system and regeneration of the forelimb of adult *Triturus*. II. Role of the sensory supply. J. Exp. Zool. *92*:297–315.
Singer, M. (1945). The nervous system and regeneration of the forelimb of adult *Triturus*. III. Role of the motor supply. J. Exp. Zool. *98*:1–21.
Singer, M. (1952). The influence of the nerve in regeneration of the amphibian extremity. Q. Rev. Biol. *27*:169–200.
Singer, M. (1954). Induction of regeneration of the forelimb of the postmetamorphic frog by augmentation of the nerve supply. J. Exp. Zool. *126*:419–471.
Singer, M. (1960). Nervous mechanisms in the regeneration of body parts in vertebrates. In *Developing Cell Systems* (D. Rudnick, ed.), Ronald Press, New York, pp. 115–133.
Singer, M. (1965). A theory of the trophic nervous control of amphibian limb regeneration, including a re-evaluation of quantitative nerve requirements. In *Regeneration in Animals and Related Problems* (V. Kiortsis and H. A. L. Trampusch, eds.), North Holland, Amsterdam, pp. 20–32.
Singer, M. (1974). Neurotrophic control of limb regeneration in the newt. Ann. N.Y. Acad. Sci. *228*:308–322.
Singer, M. (1978). On the nature of the neurotrophic phenomenon in urodele limb regeneration. Am. Zool. *18*:829–841.
Singer, M., and J. S. Andrews (1956). The adepidermal reticular network in the skin of the newt, *Triturus viridescens*. Acta Anat. *28*:313–330.
Singer, M., and J. D. Caston (1972). Neurotrophic dependence of macromolecular

synthesis in the early limb regenerate of the newt, *Triturus*. J. Embryol. Exp. Morphol. *28*:1–11.
Singer, M., and L. Craven (1948). The growth and morphogenesis of the regenerating forelimb of adult *Triturus* following denervation at various stages of development. J. Exp. Zool. *108*:279–308.
Singer, M., and F. R. L. Egloff (1949). The nervous system and regeneration of the forelimb of adult *Triturus*. VIII. The effect of limited nerve quantities on regeneration. J. Exp. Zool. *111*:295–314.
Singer, M., and J. Ilan (1977). Nerve-dependent regulation of absolute rates of protein synthesis in newt limb regenerates. Dev. Biol. *57*:174–187.
Singer, M., and M. Salpeter (1961). Regeneration in vertebrates: The role of the wound epithelium. In *Growth in Living Systems* (M. X. Zarrow, ed.), Basic Books, New York, pp. 277–311.
Singer, M., K. Rzehak, and C. E. Maier (1967). The relation between the calibre of the axon and the trophic activity of nerves in limb regeneration. J. Exp. Zool. *169*:89–98.
Singer, M., C. Maier, and W. McNutt (1976). Neurotrophic activity of brain extracts on forelimb regeneration in the urodele, *Triturus*. J. Exp. Zool. *196*:131–150.
Slack, J. M. W. (1980a). Morphogenetic properties of the skin in axolotl limb regeneration. J. Embryol. Exp. Morphol. *58*:265–288.
Slack, J. M. W. (1980b). A serial threshold theory of regeneration. J. Theor. Biol. *82*:105–140.
Slack, J. M. W. (1983). Positional information in the forelimb of the axolotl: Properties of the posterior skin. J. Embryol. Exp. Morphol. *73*:233–247.
Smith, A. R., and L. Wolpert (1975). Nerves and angiogenesis in amphibian limb regeneration. Nature (London) *257*:244–225.
Smith, A. R., J. H. Lewis, A. Crawley, and L. Wolpert (1974). A quantitative study of blastemal growth and bone regression during limb regeneration in *Triturus cristatus*. J. Embryol. Exp. Morphol. *32*:375–390.
Smith, B. H., and G. W. Kreutzberg (1976). Neuron target cell interactions. Neurosci. Res. Progr. Bull. *14*:211–453.
Smith, E. M., W. J. Meyer, and J. E. Blalock (1982). Virus-induced corticosterone in hypophysectomized mice: A possible lymphoid adrenal axis. Science *218*:1311–1312.
Smith, S. D. (1967). Inducation of partial limb regeneration in *Rana pipiens* by galvanic stimulation. Anat. Rec. *158*:89–97.
Smith, S. D. (1974). Effects of electrode placement on stimulation of adult frog limb regeneration. Ann. N.Y. Acad. Sci. *238*:500–507.
Spärck, J. V. (1969). The function of the immune response in tumor growth and its genetical regulation. Acta Path. Microbiol. Scand. 77:1–23.
Spector, G., and D. A. Willoughby (1965). Chemical mediators II. In *The Inflammatory Process* (B. W. Zweifach, L. Grant, and R. T. McCluskey, eds.), Academic, New York, pp. 389–425.
Stamatos, C., J. Squicciarini, and R. E. Fine (1983). Chick embryo spinal cord neurons synthesize a transferrin-like myotrophic protein. FEBS Lett. *153*:387–390.
Stearner, J. P. (1946) Pigmentation studies in salamanders, with special reference in the changes at metamorphosis. Physiol. Zool. *19*:375–404.
Steen, T. P. (1968). Stability of chondrocyte differentiation and contribution of muscle to cartilage during limb regeneration in the axolotl (*Siredon mexicanum*). J. Exp. Zool. *167*:49–78.
Steen, T. P. (1970). Origin and differentiative capacities of cells in the blastema of the regenerating salamander limb. Am. Zool. *10*:119–132.
Steen, T. P. (1973). The role of muscle cells of *Xenopus* limb regeneration. Am. Zool. *13*:1349–1350.

Stiles, C. D. (1983). The molecular biology of platelet-derived growth factor. Cell. *33*:653–655.

Stinson, B. D. (1964). The response of X-irradiated limbs of adult urodeles to normal tissue grafts. II. Effects of autografts of anterior or posterior halves of sixty-day forearm regenerates. J. Exp. Zool. *155*:1–24.

Stocum, D. L. (1968a). The urodele limb regeneration blastema: A self-organizing system. I. Differentiation *in vitro*. Dev. Biol. *18*:441–456.

Stocum, D. L. (1968b). The urodele limb regeneration blastema: A self-organizing system. II. Morphogenesis and differentiation of autografted whole and fractional blastemas. Dev. Biol. *18*:457–480.

Stocum, D. L. (1975a). Outgrowth and pattern formation during limb ontogeny and regeneration. Differentiation *3*:167–182.

Stocum, D. L. (1975b). Regulation after proximal or distal transposition of limb regeneration blastemas and determination of the proximal boundary of the regenerate. Dev. Biol. *45*:112–136.

Stocum, D. L. (1978). Organization of the morphogenetic field in regenerating amphibian limbs. Am. Zool. *18*:883–896.

Stocum, D. L. (1979). Stages of forelimb regeneration in *Ambystoma maculatum*. J. Exp. Zool. *209*:395–416.

Stocum, D. L. (1980a). Autonomous development of reciprocally exchanged regeneration blastemas of normal forelimbs and symmetrical hindlimbs. J. Exp. Zool. *212*:361–371.

Stocum, D. L. (1980b). Intercalary regeneration of symmetrical thighs in the axolotl, *Ambystoma mexicanum*. Dev. Biol. *79*:276–295.

Stocum, D. L. (1982). Determination of axial polarity in the urodele limb regeneration blastema. J. Embryol. Exp. Morphol. *71*:193–214.

Stocum, D. L. (1983). Morphogenesis of the amphibian limb regeneration blastema. In *Nerve, Organ and Tissue Regeneration: Research Perspectives* (F. J. Seil, ed.), Academic, New York, pp. 377–406.

Stocum, D. L. (1984). The urodele limb regeneration blastema: Determination and organization of the morphogenetic field. Differentiation *27*:13–28.

Stocum, D. L., and G. E. Dearlove (1972). Epidermal–mesodermal interaction during morphogenesis of the limb regeneration blastema in larval salamanders. J. Exp. Zool. *181*:49–61.

Stocum, D. L., and D. A. Melton (1977). Self-organizational capacity of distally transplanted regeneration blastema in larval salamanders. J. Exp. Zool. *201*:451–461.

Taban, C.-H. (1970). Contrôle immunologique de la régénération chez le triton (V). C. R. Seances Soc. Phys. Hist. Nat. (Genève) *5*:18–25.

Taban, C.-H., and P. Charollais (1965). Contrôle immunologique de la régénération (I). Note preliminaire. Arch. Sci. (Genève) *18*:103–108.

Taban, C.-H., P. Charollais, and F. Bersier (1966). Contrôle immunologique de la régénération (II). C. R. Seances Soc. Phys. Hist. Nat. (Genève) *1*:32–44.

Taban, C.-H., P. Charollais, and F. Bersier (1967a). Contrôle immunologique de la régénération (III). C. R. Seances Soc. Phys. Hist. Nat. (Genève) *1*:121–129.

Taban, C.-H., P. Charollais, and F. Bersier (1967b). Contrôle immunologique de la régénération (IV). Rev. Suisse Zool. *74*:649–663.

Taban, C.-H., M. Cathieni, and M. Schorderet (1978). Cyclic AMP and noradrenaline sensitivity fluctuations in regenerating newt tissue. Nature (London) *271*:470–472.

Takahashi, T., and M. Otsuka (1975). Regional distribution of substance P in the spinal cord and nerve roots of the cat and the effect of dorsal root section. Brain Res. *87*:1–11.

Tank, P. W. (1979). Positional information in the forelimb of the axolotl: Experiments with double-half tissues. Dev. Biol. *73*:11–24.

Tank, P. W. (1981). The ability of localized implants of whole or minced dermis to disrupt pattern formation in the regenerating forelimb of the axolotl. Am. J. Anat. *162*:315–326.

Tank, P. W., and W. Holder (1981). Pattern regulation in the regenerating limbs of urodele amphibians. Q. Rev. Biol. *56*:113–142.

Tank, P. W., B. M. Carlson, and T. G. Connelly (1976). A staging system for forelimb regeneration in the axolotl, *Ambystoma mexicanum.* J. Morphol. *150*:117–128.

Tassava, R. A. (1969). Hormonal and nutritional requirements for limb regeneration and survival of adult newts. J. Exp. Zool. *170*:33–54.

Tassava, R. A. (1983). Limb regeneration to digit stages occurs in well-fed adult newts after hypophysectomy. J. Exp. Zool. *225*:433–441.

Tassava, R. A., and M. W. Fleming, (1981). Preamputation and postamputation: Histology of the neonatal hindlimb: Implications for regeneration experiments. J. Exp. Zool. *215*:143–149.

Tassava, R. A., and D. J. Garling (1979). Regenerative responses in larval axolotl limbs with skin grafts over the amputation surface. J. Exp. Zool. *208*:97–110.

Tassava, R. A., and R. M. Loyd (1977). Injury requirement for initiation of regeneration of newt limbs which have whole skin grafts. Nature (London) *268*:49–50.

Tassava, R. A., and W. McCullough (1978). Neural control of cell cycle events in regenerating salamander limbs. Am. Zool. *18*:843–854.

Tassava, R. A., and A. L. Mescher (1975). The role of injury, nerves, and the wound epidermis during initiation of amphibian limb regeneration. Differentiation *4*:23–24.

Tassava, R. A., F. J. Chlapowski, and C. S. Thornton (1968). Limb regeneration in *Ambystoma* larvae during and after treatment with adult pituitary hormones. J. Exp. Zool. *167*:157–164.

Tassava, R. A., L. L. Bennett, and G. D. Zitnik (1974). DNA synthesis without mitosis in amputated denervated forelimbs of larval axolotls. J. Exp. Zool. *190*:111–116.

Tassava, R. A., D. P. Treece, and C. L. Olsen (1982). Effects of partial denervation on the newt blastema cell cycle. In *Limb Development and Regeneration* (A. I. Caplan and J. F. Fallon, ed.), Alan R. Liss, New York, pp. 537–546.

Tata, J. R. (1976). Growth-promoting actions of peptide hormones. In *Peptide Hormones*, Ciba Foundation Symposium 41, Elsevier, Excerpta Medica, North Holland, pp. 297–312.

Taube, E. (1921). Regeneration mit Beteiligung ortsfremder Haut bei Tritonen. Wilhelm Roux Arch. *49*:269–315.

Tepperman, J. (1965). *Metabolic and Endocrine Physiology*, Year Book Medical, Chicago, pp. 145–167.

Thom, R. (1975). *Structural Stability and Morphogenesis*, Benjamin, Reading Mass.

Thomas, K. A., M. C. Riley, S. K. Lemmon, N. C. Baglan, and R. A. Bradshaw (1980). Brain fibroblast growth factor: Nonidentity with myelin basic protein fragments. J. Biol. Chem. *255*:5517–5520.

Thomas, K. A., M. Riss-Candelore, and S. Fitzpatrick (1984). Purification and characterization of acidic fibroblast growth factor from bovine brain. Proc. Nat. Acad. Sci. (U.S.A.) *81*:357–361.

Thornton, C. S. (1938). The histogenesis of muscle in the regenerating forelimb of larval *Amblystoma punctatum.* J. Morphol. *62*:17–47.

Thornton, C. S. (1942). Studies on the origin of the regeneration blastema in *Triturus viridescens.* J. Exp. Zool. *89*:375–389.

Thornton, C. S. (1958). The inhibition of limb regeneration in urodele larvae by localized irradiation with ultraviolet light. J. Exp. Zool. *137*:153–180.

Thornton, C. S. (1960). Influence of an eccentric epidermal cap on limb regeneration in *Amblystoma* larvae. Dev. Biol. *2*:551–569.

Thornton, C. S. (1962). Eccentric blastema formation in aneurogenic limbs of *Amblystoma* larvae following epidermal cap deviation. Dev. Biol. *5*:328–343.

Thornton, C. S. (1968). Amphibian limb regeneration. Adv. Morphog. 7:205–250.

Thornton, C. S., and M. T. Thornton (1965). The regeneration of accessory limb parts following epidermal cap transplantation in urodeles. Experientia *21*:1–6.

Tickle, C. (1981). Limb regeneration. Am. Sci. *69*:639–646.

Tilney, L. G. (1980). Actin, motility, and membranes. In *Society of General Physiology Series: Membrane Transduction Mechanisms* Vol. 33 (R. A. Cone and J. E. Dowling, eds.), Raven Press, New York, pp. 163–186.

Tomlinson, B. L., M. Globus, and S. Vethamany-Globus (1981). Promotion of mitosis in cultured newt limb regenerates by a diffusible nerve factor. In Vitro *17*:167–172.

Tomlinson, B. L., M. Globus, and S. Vethamany-Globus (1982). Studies of mitosis in excised limb regenerates of the newt, *Notophthalmus viridescens*. J. Exp. Zool. *223*:115–122.

Tomlinson, B. L., M. Globus, and S. Vethamany-Globus (1984). The blastema cell cycle *in vitro* and attempted restoration of blastema cell cycling in denervated blastemata of the adult newt, *Notophthalmus viridescens*. J. Exp. Zool. *232*:249–258.

Toole, B. P., and J. Gross (1971). The extracellular matrix of the regenerating newt limb: Synthesis and removal of hyaluronate prior to differentiation. Dev. Biol. *25*:57–77.

Tornier, G. (1906). Kampf der Gewebe im Regenerat bei Begunstigung der Hautregeneration. Wilhelm Roux Arch. *22*:348–369.

Trampusch, H. A. L. (1958a). The action of X-rays on the morphogenetic field. I. Heterotopic grafts on irradiated limb. Ned. Akad. Weten. Proc. K Ser. C *61*:417–430.

Trampusch, H. A. L. (1958b). The action of X-rays on the morphogenetic field. II. Heterotopic skin on irradiated tails. Ned. Akad. Weten. Proc. K Ser. C *61*:530–545.

Trampusch, H. A. L. (1959). The effect of X-rays on regenerative capacity. In *Regeneration* in Vertebrates (C. S. Thornton, ed.), University of Chicago Press, Chicago, pp. 83–97.

Trampusch, H. A. L., and A. E. Harrebomée (1965). Dedifferentiation, a prerequisite of regeneration. In *Regeneration in Animals and Related Problems* (V. Kiortisis and H. A. L. Trampusch, eds.), North Holland, Amsterdam, pp. 341–474.

Trampusch, H. A. L., and A. E. Harrebomée (1969). Dedifferentiation and the interconvertibility of the different cell-types in the amphibian extremity. Acta Embryol. Exp. *1969*:35–69.

Trowell, O. E., and E. N. Willmer (1939). Studies on the growth of tissues *in vitro*. VI. The effects of some tissue extracts on the growth of periosteal fibroblasts. J. Exp. Biol. *16*:60–82.

Turner, C. D. (1966). *General Endocrinology*, Saunders, Philadelphia.

Umanski, E. (1938). The regeneration potencies of axolotl skin studied by means of exclusion of the regeneration capacity of tissues through exposure to X-rays. Bull. Biol. Med. Exp. (URSS) *6*:141–145.

Ussing, H. H. (1964). Transport of electrolytes and water epithelia. Harvey Lect. *59*:1–30.

Vanable, J. W., Jr., L. L. Hearson, and M. E. McGinnis (1983). The role of endogenous electrical fields in limb regeneration. In *Limb Development and Regeneration* (John F. Fallon and A. I. Caplan, eds.), Alan R. Liss, New York, pp. 587–596.

Van Arsdall, C. B., and T. L. Lentz (1968). Neurons: Secretory activity during limb regeneration and induction in the newt. Science *162*:1296–1297.

Verge, P., L. Hakansson, and R. Hollgren (1982). The effect of hyaluronic acid on neutrophil function *in vitro* and *in vivo*. In *Biochemistry and Function of Phagocytes* (F. Rossi and P. Patriarca eds.), Plenum, New York, pp. 559–564.

Vethamany, S. (1970). *In Vivo* and *In Vitro* Studies on the Influence of Hormones on Limb and Tail Regeneration in Adult *Diemictylus viridescens*. Ph.D. thesis, University of Toronto, Toronto, Canada.

Vethamany-Globus, S., and R. A. Liversage (1973a). *In vitro* studies of the influence of hormones on tail regeneration in adult *Diemictylus viridescens*. J. Embryol. Exp. Morphol. *30*:397–413.

Vethamany-Globus, S., and R. A. Liversage (1973b). The relationship between the anterior pituitary gland and the pancreas in tail regeneration of the adult newt. J. Embryol. Exp. Morphol. *30*:415–426.

Vethamany-Globus, S., and R. A. Liversage (1973c). Effects of insulin insufficiency on forelimb and tail regeneration in adult *Diemictylus viridescens*. J. Embryol. Exp. Morphol. *30*:427–447.

Vethamany-Globus, S., M. Globus, J. Harford, I. Fraser, and D. Weber (1977). Hormone control in regeneration: Effects of somatostatin on appendage regeneration, blood glucose and liver glycogen in *Diemictylus viridescens*. J. Embryol. Exp. Morphol. *40*:115–124.

Vethamany-Globus, S., M. Globus, and B. Tomlinson (1978). Neural and hormonal stimulation of DNA and protein synthesis in cultured regeneration blastemata in the newt *Notophthalmus viridescens*. Dev. Biol. *65*:183–192.

Vethamany-Globus, S., M. Globus, A. Darch, G. Milton, B. L. Tomlinson, and D. Roth (1984). *In vitro* effects of macromolecular synthetic events in newt limb regeneration blastemata. J. Exp. Zool. *231*:63–74.

Von Euler, U. S., and J. H. Gaddum (1931). An unidentified depressor substance in certain tissue extracts. J. Physiol. *72*:74.

Von Haam, E., and L. Cappel (1940). Effects of hormones upon cells grown *in vitro*. Am. J. Cancer *39*:354–359.

Vorontsova, M. A., and L. D. Liosner (1960). *Asexual Propagation and Regeneration*, Pergamon, London.

Wahl, S. M., and L. H. Wahl (1979). Lymphokine modulation of connective tissue metabolism. Ann. N.Y. Acad. Sci. *332*:411–422.

Wallace, H. (1981). *Vertebrate Limb Regeneration*, Wiley, New York.

Wallace, H., and M. Maden (1976). The cell cycle during amphibian limb regeneration. J. Cell Sci. *20*:539–548.

Wallace, H., S. Wessels, and H. Conn (1971). Radioresistance of nerves in amphibian limb regeneration. Wilhelm Roux Arch. *166*:219–225.

Wallace, H., B. M. Wallace, and M. Maden (1974). Participation of cartilage grafts in amphibian limb regeneration. J. Embryol. Exp. Morphol. *32*:391–404.

Walter, F. K. (1910). Schilddrüse und Regeneration. Wilhelm Roux Arch. *31*:91–103.

Ward, P. A. (1980). Inflammation. In *Principles of Pathobiology* (R. B. Hill and M. F. LaVia, eds.), Oxford University Press, New York, pp. 112–162.

Webster, G., and B. Goodwin (1981). History and structure in biology. Perspect. Biol. Med. *25*:39–62.

Weiss, B., and T. L. Wallace (1980). Mechanisms and pharmacological implications of altering calmodulin activity. In *Calcium and Cell Function* (W. Y. Cheung, ed.), Academic, New York, p. 329.

Weis, P. (1926). Ganzregenerate aus halben Extremitätenquerschnitt. Wilhelm Roux Arch. *107*:1–54.

Wertz, R. L., and D. J. Donaldson (1980). Early events following amputation of adult newt limbs given regeneration inhibitory doses of X-irradiation. Dev. Biol. *74*:434–445.

Westall, F. C., V. A. Lennon, and D. Gospodarowicz (1978). Brain-derived fibroblast growth factor: Identity with a fragment of the basic protein of myelin. Proc. Nat. Acad. Sci. (U.S.A.) *75*:4675–4678.

Whiteside, C. (1978). Partial Denervation Effects on Cell Cycle Parameters During Limb Regeneration in Axolotls. M.Sc. thesis, Ohio State University, Columbus, Ohio.

Whitfield, J. F., A. L. Boynton, J. P. Macmanus, M. Sikorska, and B. K. Tsang (1979). The regulation of cell proliferation by calcium and cyclic AMP. Mol. Cell. Biochem. *27*:155–179.

Whittaker, V. P., I. A. Michaelson, and R. J. Kirkland (1964). The separation of synaptic vesicles from nerve-ending particles ("synaptosomes"). Biochem. J. *90*:293–303.

Wilhelm, D. L. (1965). Chemical mediators I. In *The Inflammatory Process* (B. W. Zweifach, L. Grant, and R. T. McCluskey, eds.), Academic, New York, pp. 389–425.

Wilkerson, J. A. (1963). The role of growth hormone in regeneration of the forelimb of the hypophysectomized newt. J. Exp. Zool. *154*:223–230.

Wolpert, L. (1969). Positional information and the spatial pattern of cellular differentiation. J. Theor. Biol. *25*:1–47.

Wolpert, L. (1971). Positional information and pattern formation. Curr. Top. Dev. Biol. 6:183–224.

Woodruff, R. I., and W. H. Telfer (1980). Electrophoresis of proteins in intercellular bridges. Nature (London) *84*:286–288.

Wurster, D. H., and M. R. Miller (1960). Studies on the blood glucose and pancreatic islets of the salamander, *Taricha torosa.* Comp. Biochem. Physiol. *1*:101–109.

Yachnis, A. T., and A. L. Mescher (1982). Stimulation of DNA synthesis in Balb/c 3T3 cells by peripheral nerve degenerating *in vitro.* Exp. Neurol. 76:139–149.

Yaksh, T. L., T. M. Jessel, R. Gamse, A. W. Mudge, and S. E. Leeman (1980). Intrathecal morphine inhibits substance P release from mammalian spinal cord *in vivo.* Nature (London) *286*:155–157.

Yaoi, Y., and T. Kanaseki (1972). Role of microexudate carpet in cell division. Nature (London) *237*:283–285.

Yoshimoto, Y., A. R. Wolfsen, and W. D. Odell (1977). Human chorionic gonadotropin-like substance in nonendocrine tissues of normal subjects. Science *197*:575–577.

Index

Printed and bound in the UK by
CPI Antony Rowe, Eastbourne

The manufacturer's authorised representative in the EU for product safety is Oxford University Press España S.A. of El Parque Empresarial San Fernando de Henares, Avenida de Castilla, 2 - 28830 Madrid (www.oup.es/en or product.safety@oup.com). OUP España S.A. also acts as importer into Spain of products made by the manufacturer.
Printed and bound by CPI Group (UK) Ltd, Croydon, CR0 4YY
12/09/2026
02215930-0005